An Introduction to Extremal Kähler Metrics

An Introduction to Extremal Kähler Metrics

Gábor Székelyhidi

Graduate Studies in Mathematics

Volume 152

American Mathematical Society
Providence, Rhode Island

2010 *Mathematics Subject Classification.* Primary 53C25, 53C55, 14L24.

For additional information and updates on this book, visit
www.ams.org/bookpages/gsm-152

Library of Congress Cataloging-in-Publication Data

Székelyhidi, Gábor, 1981–
An introduction to extremal Kähler metrics / Gábor Székelyhidi.
pages cm. — (Graduate studies in mathematics ; volume 152)
Includes bibliographical references and index.
ISBN 978-1-4704-1047-6 (alk. paper)
1. Geometry, Differential. I. Title.

QA641.S94 2014
516.3′73—dc23

2014006619

∞ The paper used in this book is acid-free and falls within the guidelines
established to ensure permanence and durability.
Visit the AMS home page at http://www.ams.org/

10 9 8 7 6 5 4 3 2 1 19 18 17 16 15 14

To Sonja and Nóra

Contents

Preface

This book grew out of lecture notes written for a graduate topics course taught at the University of Notre Dame in the spring of 2012. The goal is to quickly introduce graduate students to ideas surrounding recent developments on extremal Kähler metrics. We make an effort to introduce the main ideas from Kähler geometry and analysis that are required, but the parts of the book on geometric invariant theory and K-stability would be difficult to follow without more background in complex algebraic geometry. A reader with a background in Riemannian geometry and graduate level analysis should be able to follow the rest of the book.

I would like to thank Tamás Darvas, Yueh-Ju Lin, and Valentino Tosatti for helpful comments about the manuscript.

Gábor Székelyhidi
Notre Dame, 2014

Introduction

A basic problem in differential geometry is to find canonical, or best, metrics on a given manifold. There are many different incarnations of this, perhaps the most well known being the classical uniformization theorem for Riemann surfaces. The study of extremal metrics is an attempt at finding a higher-dimensional generalization of this result in the setting of Kähler geometry. Extremal metrics were introduced by Calabi in the 1980s as an attempt to find canonical Kähler metrics on Kähler manifolds as critical points of a natural energy functional. The energy functional is simply the L^2-norm of the curvature of a metric. The most important examples of extremal metrics are Kähler-Einstein metrics and constant scalar curvature Kähler (or cscK) metrics.

It turns out that extremal metrics do not always exist, and the question of their existence is particularly interesting on projective manifolds. In this case, by works of Yau, Tian, and Donaldson, it was realized that the existence of extremal metrics is related to the stability of the manifold in an algebro-geometric sense, and obtaining a necessary and sufficient condition of this form for existence is the central problem in the field. Our goal in this book is to introduce the reader to some of the basic ideas on both the analytic and the algebraic sides of this problem. One concrete goal is to give a fairly complete proof of the following result.

Theorem. *If M admits a cscK metric in $c_1(L)$ for an ample line bundle $L \to M$ and if M has no non-trivial holomorphic vector fields, then the pair (M, L) is K-stable.*

The converse of this result, i.e. the existence of cscK metrics on K-stable manifolds, is the central conjecture in the field.

We will now give a brief description of the contents of the book. The first two chapters give a quick review of some of the background material that is needed. The first chapter contains the basic definitions in Kähler geometry, with a focus on calculations in local coordinates. The second chapter focuses on some of the analytic background required, in particular the Schauder estimates for elliptic operators, which we prove using a blow-up argument due to L. Simon.

The topic of Chapter 3 is Kähler-Einstein metrics, which are a special case of extremal metrics. We give a proof of Yau's celebrated theorem on the solution of the complex Monge-Ampère equation, leading to existence results for Kähler-Einstein metrics with zero or negative Ricci curvature. The case of positive Ricci curvature has only been understood very recently through the work of Chen, Donaldson, and Sun. The details of this are beyond the scope of this book, and we only give a very brief discussion in Section 3.5.

The study of general extremal metrics begins in Chapter 4. Following Calabi, we introduce extremal metrics as critical points of the Calabi functional, which is the L^2-norm of the scalar curvature:

$$\omega \mapsto \int_M S(\omega)^2\,\omega^n,$$

defined for metrics ω in a fixed Kähler class. An important discovery is that extremal metrics have an alternative variational characterization, as critical points of the (modified) Mabuchi functional. This is convex along geodesics in the space of Kähler metrics with respect to a natural, infinite-dimensional, Riemannian structure. Moreover the variation of the Mabuchi functional is closely related to the Futaki invariant, which plays a prominent role in the definition of K-stability. After giving the basic definitions, we construct an explicit family of extremal metrics on a ruled surface due to Tønnesen-Friedman in Section 4.4. This example illustrates how a sequence of extremal metrics can degenerate, and we return to it again in Section 6.5. In Section 4.5 we give an introduction to the study of extremal metrics on toric manifolds. Toric manifolds provide a very useful setting in which to study extremal metrics and stability, and while in the two-dimensional case the basic existence question is understood through the works of Donaldson and of Chen, Li, and Sheng, the higher-dimensional case remains an important problem to study.

In Chapter 5 we give an introduction to the relation between symplectic and algebraic quotients—the Kempf-Ness theorem—which, at least on a heuristic level, underpins many of the ideas that have to do with extremal metrics. The general setting is a compact group K acting by Hamiltonian isometries on a Kähler manifold M, with a moment map $\mu : M \to \mathfrak{k}^*$. The

Kempf-Ness theorem characterizes those orbits of the complexified group K^c which contain zeros of the moment map. The reason why this is relevant is that the scalar curvature of a Kähler metric, or rather the map $\omega \mapsto S(\omega) - \hat{S}$ where $\hat{S}$ is the average scalar curvature, can be realized as a moment map for a suitable infinite-dimensional Hamiltonian action. At the same time, orbits of K^c can be thought of as metrics in a given Kähler class, so an infinite-dimensional analog of the Kempf-Ness theorem would describe Kähler classes that contain cscK metrics. In Section 5.5 we will describe a suitable extension of the Kempf-Ness theorem dealing with critical points of the norm squared of a moment map, which in the infinite-dimensional setting are simply extremal metrics.

The notion of K-stability is studied in Chapter 6. It is defined in analogy with the Hilbert-Mumford criterion in geometric invariant theory by requiring that a certain weight—the Donaldson-Futaki invariant—is positive for all $\mathbf{C}^*$-equivariant degenerations of the manifold. These degenerations are called test-configurations. In analogy with the finite-dimensional setting of the Kempf-Ness theorem, the Donaldson-Futaki invariant of a test-configuration can be seen as an attempt at encoding the asymptotics of the Mabuchi functional "at infinity", with the positivity of the weights ensuring that the functional is proper. In Section 6.6 we will describe test-configurations from the point of view of filtrations of the homogeneous coordinate ring of the manifold. It is likely that the notion of K-stability needs to be strengthened to ensure the existence of a cscK metric, and filtrations allow for a natural way to enlarge the class of degenerations that we consider. In the case of toric varieties, passing from test-configurations to filtrations amounts to passing from rational piecewise linear convex functions to all continuous convex functions, as we will discuss in Section 6.7.

The basic tool in relating the differential geometric and algebraic aspects of the problem is the Bergman kernel, which we discuss in Chapter 7. We first give a proof of a simple version of the asymptotic expansion of the Bergman kernel going back to Tian, based on the idea of constructing peaked sections of a sufficiently high power of a positive line bundle. Then, following Donaldson, we use this to show that a projective manifold which admits a cscK metric must be K-semistable. This is a weaker statement than the theorem stated above. The Bergman kernel also plays a key role in the recent developments on Kähler-Einstein metrics, through the partial C^0-estimate conjectured by Tian. We will discuss this briefly in Section 7.6.

In the final chapter, Chapter 8, the main result is a perturbative existence result for cscK metrics due to Arezzo and Pacard. Starting with a cscK metric ω on M and assuming that M has no non-zero holomorphic vector fields, we show that the blow-up of M at any point admits cscK metrics in suitable Kähler classes. The gluing technique used together with analysis in weighted Hölder spaces has many applications in geometric analysis. Apart from giving many new examples of cscK manifolds, this existence result is crucial in the final step of proving the theorem stated above, namely to improve the conclusion from K-semistability (obtained in Chapter 7) to K-stability. The idea due to Stoppa is to show that if M admits a cscK metric and is not K-stable, then a suitable blow-up of M is not even K-semistable. Since the blow-up admits a cscK metric, this is a contradiction.

There are several important topics that are missing from this book. We make almost no mention of parabolic equations such as the Calabi flow and the Kähler-Ricci flow. We also do not discuss in detail the existence theory for constant scalar curvature metrics on toric surfaces and for Kähler-Einstein metrics on Fano manifolds since each of these topics could take up an entire book. It is our hope that after studying this book the reader will be eager and ready to tackle these more advanced topics.

Chapter 1

Kähler Geometry

In this chapter we cover some of the background from Kähler geometry that we will need. Rather than formally setting up the theory, we will focus on how to do calculations with covariant derivatives and the curvature tensor on Kähler manifolds in local coordinates. For a much more thorough treatment of the subject good references are Griffiths-Harris [**59**] and Demailly [**39**].

1.1. Complex manifolds

A complex manifold M can be thought of as a smooth manifold on which we have a well-defined notion of holomorphic function. More precisely, for an integer $n > 0$ (the complex dimension), M is covered by open sets U_α, together with homeomorphisms

$$\varphi_\alpha : U_\alpha \to V_\alpha \subset \mathbf{C}^n,$$

such that the "transition maps" $\varphi_\alpha \circ \varphi_\beta^{-1}$ are holomorphic wherever they are defined. A function $f : M \to \mathbf{C}$ is then holomorphic if the composition $f \circ \varphi_\alpha^{-1}$ is holomorphic on V_α, for all α. Using these charts, near any point $p \in M$ there exists a holomorphic coordinate system $z^1, \ldots, z^n$, consisting of complex-valued functions with $z^i(p) = 0$ for each i. Moreover if $w^1, \ldots, w^n$ form a different holomorphic coordinate system, then each w^i is a holomorphic function of the $z^1, \ldots, z^n$.

Example 1.1 (The Riemann sphere). We let $M = S^2$, and we think of $S^2 \subset \mathbf{R}^3$ as the unit sphere. Identify the xy-plane in $\mathbf{R}^3$ with $\mathbf{C}$. We define two charts. Let U_1 be the complement of the "north pole", i.e. $U_1 = S^2 \setminus \{(0, 0, 1)\}$, and define

$$\varphi : U_1 \to \mathbf{C}$$

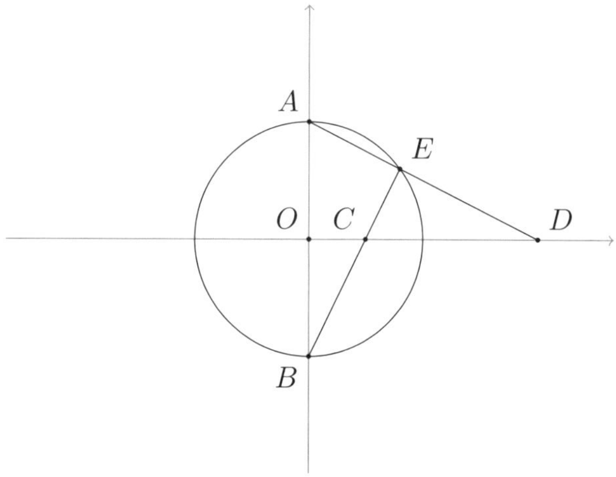

Figure 1.1. A cross-section of the stereographic projections.

to be the stereographic projection from the north pole to the xy-plane. Similarly let $U_2 = S^2 \setminus \{(0,0,-1)\}$ be the complement of the south pole, and let

$$\psi : U_2 \to \mathbf{C}$$

be the composition of the stereographic projection to the xy-plane from the south pole, with complex conjugation. One can then compute that

$$\psi \circ \varphi^{-1}(z) = \frac{1}{z} \quad \text{for } z \in \mathbf{C} \setminus \{0\}. \tag{1.1}$$

Indeed, in Figure 1.1 the points C and D are the stereographic projections of E from the south and north poles, respectively. The triangles OBC and ODA are similar, from which it follows that $|OC| \cdot |OD| = 1$, i.e. the two stereographic projections are related by inversion in the unit circle. Inversion in the unit circle is the transformation $z \mapsto \overline{z}^{-1}$, so when we compose with complex conjugation, we obtain the transition function (1.1).

Since this transition function is holomorphic, our two charts give S^2 the structure of a complex manifold. Note that if we do not compose the projection with complex conjugation when defining ψ, then even the orientations defined by φ and ψ would not match, although the two charts would still give S^2 the structure of a smooth manifold.

Example 1.2 (Complex projective space)**.** The complex projective space $\mathbf{CP}^n$ is defined to be the space of complex lines in $\mathbf{C}^{n+1}$. In other words, points of $\mathbf{CP}^n$ are $(n+1)$-tuples $[Z_0 : \cdots : Z_n]$, where not every entry is zero, and we identify

$$[Z_0 : \cdots : Z_n] = [\lambda Z_0 : \cdots : \lambda Z_n]$$

for all $\lambda \in \mathbf{C} \setminus \{0\}$. As a topological space, $\mathbf{CP}^n$ inherits the quotient topology from $\mathbf{C}^{n+1} \setminus \{(0,\ldots,0)\}$ under this equivalence relation. We call the $Z_0,\ldots,Z_n$ homogeneous coordinates. To define the complex structure we will use $n+1$ charts. For $i \in \{0,1,\ldots,n\}$, let

$$U_i = \left\{[Z_0 : \ldots : Z_n] \,\middle|\, Z_i \neq 0\right\}$$

and

$$\varphi_i : U_i \to \mathbf{C}^n$$

$$[Z_0 : \ldots : Z_n] \mapsto \left(\frac{Z_0}{Z_i}, \ldots, \widehat{\frac{Z_i}{Z_i}}, \ldots, \frac{Z_n}{Z_i}\right),$$

where the $\dfrac{Z_i}{Z_i}$ term is omitted. It is then easy to check that the transition functions are holomorphic. For example, using coordinates $w^1,\ldots,w^n$ on $\mathbf{C}^n$ we have

$$\varphi_1 \circ \varphi_0^{-1}(w^1,\ldots,w^n) = \left(\frac{1}{w^1}, \frac{w^2}{w^1}, \ldots, \frac{w^n}{w^1}\right). \tag{1.2}$$

In the case $n=1$ we obtain two charts with the same transition function as in the previous example, so $\mathbf{CP}^1 = S^2$ as complex manifolds.

Topologically $\mathbf{CP}^n$ can be seen as a quotient S^{2n+1}/S^1, where $S^{2n+1} \subset \mathbf{C}^{n+1}$ is the unit sphere, and S^1 acts as multiplication by unit length complex numbers. It follows that $\mathbf{CP}^n$ is compact.

Example 1.3 (Projective manifolds)**.** Suppose that $f_1,\ldots,f_k$ are homogeneous polynomials in $Z_0,\ldots,Z_n$. Even though the f_i are not well-defined functions on $\mathbf{CP}^n$ (we will later see that they are sections of line bundles), their zero sets are well-defined. Let $V \subset \mathbf{CP}^n$ be their common zero set

$$V = \left\{[Z_0 : \ldots : Z_n] \,\middle|\, f_i(Z_0,\ldots,Z_n) = 0 \text{ for } i = 1,\ldots,k\right\}.$$

If V is a smooth submanifold, then it is a complex manifold and charts can be constructed using the implicit function theorem. Being closed subsets of a compact space, projective manifolds are compact.

These projective manifolds are general enough that in this book they are essentially the only complex manifolds with which we will be concerned. They lie at the intersection of complex differential geometry and algebraic geometry and we will require tools from both fields. In particular the basic question we will ask is differential geometric in nature, about the existence of certain special metrics on projective manifolds. In studying this question, however, one is naturally led to consider the behavior of projective manifolds in families and their degenerations to possibly singular limiting spaces. Algebraic geometry will provide a powerful tool to study such problems.

1.2. Almost complex structures

An alternative way to introduce complex manifolds is through almost complex structures.

Definition 1.4. An almost complex structure on a smooth manifold M is an endomorphism $J : TM \to TM$ of the tangent bundle such that $J^2 = -\mathrm{Id}$, where Id is the identity map.

In other words an almost complex structure equips the tangent space at each point with a linear map which behaves like multiplication by $\sqrt{-1}$. The dimension of M must then be even, since any endomorphism of an odd-dimensional vector space has a real eigenvalue, which could not square to -1.

Example 1.5. If M is a complex manifold, then the holomorphic charts identify each tangent space T_pM with $\mathbf{C}^n$, so we can define $J(v) = \sqrt{-1}v$ for $v \in T_pM$, giving an almost complex structure. The fact that the transition functions are holomorphic means precisely that multiplication by $\sqrt{-1}$ is compatible under the different identifications of T_pM with $\mathbf{C}^n$ using different charts.

If $z^1, \ldots, z^n$ are holomorphic coordinates and $z^i = x^i + \sqrt{-1}y^i$ for real functions x^i, y^i, then we can also write

$$J\left(\frac{\partial}{\partial x^i}\right) = \frac{\partial}{\partial y^i}, \quad J\left(\frac{\partial}{\partial y^i}\right) = -\frac{\partial}{\partial x^i}.$$

Definition 1.6. An almost complex structure is called integrable if it arises from holomorphic charts as in the previous example. We will use the term "complex structure" to mean an integrable almost complex structure.

On complex manifolds it is convenient to work with the complexified tangent bundle

$$T^{\mathbf{C}}M = TM \otimes_{\mathbf{R}} \mathbf{C}.$$

In terms of local holomorphic coordinates it is convenient to use the basis

$$\left\{\frac{\partial}{\partial z^1}, \ldots, \frac{\partial}{\partial z^n}, \frac{\partial}{\partial \overline{z}^1}, \ldots, \frac{\partial}{\partial \overline{z}^n}\right\}, \tag{1.3}$$

where in terms of the real and imaginary parts $z^i = x^i + \sqrt{-1}y^i$ we have

$$\frac{\partial}{\partial z^i} = \frac{1}{2}\left(\frac{\partial}{\partial x^i} - \sqrt{-1}\frac{\partial}{\partial y^i}\right) \quad \text{and} \quad \frac{\partial}{\partial \overline{z}^i} = \frac{1}{2}\left(\frac{\partial}{\partial x^i} + \sqrt{-1}\frac{\partial}{\partial y^i}\right). \tag{1.4}$$

The endomorphism J extends to a complex linear endomorphism of $T^{\mathbf{C}}M$ and induces a decomposition of this bundle pointwise into the $\sqrt{-1}$ and $-\sqrt{-1}$ eigenspaces

$$T^{\mathbf{C}}M = T^{1,0}M \oplus T^{0,1}M.$$

In terms of local holomorphic coordinates, $T^{1,0}M$ is spanned by the $\dfrac{\partial}{\partial z^i}$ while $T^{0,1}M$ is spanned by the $\dfrac{\partial}{\partial \overline{z}^i}$.

Similarly we can complexify the cotangent bundle to obtain $\Omega^1_{\mathbf{C}}M$, which is decomposed according to the eigenvalues of the endomorphism dual to J (which we will still denote by J) into

$$\Omega^1_{\mathbf{C}}M = \Omega^{1,0}M \oplus \Omega^{0,1}M.$$

In terms of coordinates, $\Omega^{1,0}$ is spanned by $dz^1, \ldots, dz^n$, while $\Omega^{0,1}$ is spanned by $d\overline{z}^1, \ldots, d\overline{z}^n$, where

$$dz^i = dx^i + \sqrt{-1}dy^i \quad \text{and} \quad d\overline{z}^i = dx^i - \sqrt{-1}dy^i.$$

Moreover $\{dz^1, \ldots, dz^n, d\overline{z}^1, \ldots, d\overline{z}^n\}$ is the basis dual to (1.3).

The decomposition extends to higher-degree forms

$$\Omega^r_{\mathbf{C}}M = \bigoplus_{p+q=r} \Omega^{p,q}M,$$

where $\Omega^{p,q}M$ is locally spanned by

$$dz^{i_1} \wedge \cdots \wedge dz^{i_p} \wedge d\overline{z}^{j_1} \wedge \cdots \wedge d\overline{z}^{j_q}.$$

On a complex manifold the decomposition of forms gives rise to a decomposition of the exterior derivative as $d = \partial + \overline{\partial}$, where

$$\partial : \Omega^{p,q}M \to \Omega^{p+1,q}M,$$
$$\overline{\partial} : \Omega^{p,q}M \to \Omega^{p,q+1}M$$

are two projections of d. A useful observation is that $\overline{\partial}\alpha = \overline{\partial\overline{\alpha}}$ for any form α.

Example 1.7. A function $f : M \to \mathbf{C}$ is holomorphic if and only if $\overline{\partial} f = 0$, since

$$\overline{\partial} f = \frac{\partial f}{\partial \overline{z}^1} d\overline{z}^1 + \cdots + \frac{\partial f}{\partial \overline{z}^n} d\overline{z}^n,$$

and $\dfrac{\partial f}{\partial \overline{z}^i}$ are the Cauchy-Riemann equations.

Example 1.8. For a function $f : M \to \mathbf{R}$, the form $\sqrt{-1}\partial\overline{\partial} f$ is a real $(1,1)$-form, a kind of complex Hessian of f. In particular if $f : \mathbf{C} \to \mathbf{R}$, then

$$\begin{aligned}
\sqrt{-1}\partial\overline{\partial} f &= \sqrt{-1}\left(\frac{\partial}{\partial z}\frac{\partial f}{\partial \overline{z}}\right) dz \wedge d\overline{z} \\
&= \frac{\sqrt{-1}}{4}\left(\frac{\partial}{\partial x} - \sqrt{-1}\frac{\partial}{\partial y}\right)\left(\frac{\partial f}{\partial x} + \sqrt{-1}\frac{\partial f}{\partial y}\right)(dx + \sqrt{-1}dy) \wedge (dx - \sqrt{-1}dy) \\
&= \frac{1}{2}\left(\frac{\partial^2 f}{\partial x^2} + \frac{\partial^2 f}{\partial y^2}\right) dx \wedge dy.
\end{aligned}$$

1.3. Hermitian and Kähler metrics

Let M be a complex manifold with complex structure J. We will be interested in Riemannian metrics on M which are compatible with the complex structure in a particularly nice way. Recall that a Riemannian metric is a positive definite symmetric bilinear form on each tangent space.

Definition 1.9. A Riemannian metric g on M is Hermitian if $g(JX, JY) = g(X, Y)$ for any tangent vectors X, Y. In other words we require J to be an orthogonal transformation on each tangent space.

Given a Hermitian metric g, we define $\omega(X, Y) = g(JX, Y)$ for any X, Y. Then ω is antisymmetric in X, Y and one can check that in this way ω defines a real 2-form of type $(1, 1)$.

Definition 1.10. A Hermitian metric g is Kähler if the associated 2-form ω is closed, i.e. $d\omega = 0$. Then ω is called the Kähler form, but often we will call ω the Kähler metric and make no mention of g.

In local coordinates $z^1, \ldots, z^n$ a Hermitian metric is determined by the components $g_{j\bar{k}}$ where

$$g_{j\bar{k}} = g\left(\frac{\partial}{\partial z^j}, \frac{\partial}{\partial \bar{z}^k}\right),$$

and we are extending g to complex tangent vectors by complex linearity in both entries. The Hermitian condition implies that for any j, k we have

$$g\left(\frac{\partial}{\partial z^j}, \frac{\partial}{\partial z^k}\right) = g\left(\frac{\partial}{\partial \bar{z}^j}, \frac{\partial}{\partial \bar{z}^k}\right) = 0.$$

In terms of the components $g_{j\bar{k}}$ we can therefore write

$$g = \sum_{j,k} g_{j\bar{k}}(dz^j \otimes d\bar{z}^k + d\bar{z}^k \otimes dz^j).$$

Note that the bar on $\bar{k}$ in the components $g_{j\bar{k}}$ is used to remember the distinction between holomorphic and antiholomorphic components.

The symmetry of g implies that $\overline{g_{j\bar{k}}} = g_{k\bar{j}}$, and the positivity of g means that $g_{j\bar{k}}$ is a positive definite Hermitian matrix at each point. The associated 2-form ω can be written as

$$\omega = \sqrt{-1} \sum_{j,k} g_{j\bar{k}} dz^j \wedge d\bar{z}^k,$$

and finally g is Kähler if for all i, j, k we have

$$\frac{\partial}{\partial z^i} g_{j\bar{k}} = \frac{\partial}{\partial z^j} g_{i\bar{k}}.$$

Exercise 1.11. Show that on a Kähler manifold (M, ω) of complex dimension n, the Riemannian volume form is given by $\dfrac{\omega^n}{n!}$, where $\omega^n = \omega \wedge \cdots \wedge \omega$.

Example 1.12 (Fubini-Study metric)**.** The complex projective space $\mathbf{CP}^n$ has a natural Kähler metric ω_{FS} called the Fubini-Study metric. To construct it, recall the projection map $\pi : \mathbf{C}^{n+1} \setminus \{0\} \to \mathbf{CP}^n$. A section s over an open set $U \subset \mathbf{CP}^n$ is a holomorphic map $s : U \to \mathbf{C}^{n+1}$ such that $\pi \circ s$ is the identity. Given such a section, we define

$$\omega_{FS} = \sqrt{-1}\partial\bar{\partial} \log \|s\|^2.$$

To check that this is well-defined, note that if s' is another section over an open set V, then on the intersection $U \cap V$ we have $s' = fs$ for a holomorphic function $f : U \cap V \to \mathbf{C} \setminus \{0\}$ and

$$\begin{aligned}\sqrt{-1}\partial\bar{\partial} \log \|fs\|^2 &= \sqrt{-1}\partial\bar{\partial} \log \|s\|^2 + \sqrt{-1}\partial\bar{\partial} \log f + \sqrt{-1}\partial\bar{\partial} \log \overline{f} \\ &= \sqrt{-1}\partial\bar{\partial} \log \|s\|^2.\end{aligned}$$

Since sections exist over small open sets U, we obtain a well-defined, closed $(1,1)$-form on $\mathbf{CP}^n$. The form ω_{FS} is $U(n+1)$-invariant, and $U(n+1)$ acts transitively on $\mathbf{CP}^n$ so it is enough to check that the corresponding Hermitian matrix is positive definite at a single point. At the point $[1 : 0 : \cdots : 0]$ let us use local holomorphic coordinates

$$z^i = \frac{Z_i}{Z_0} \quad \text{for } i = 1, \ldots, n$$

on the chart U_0. A section is then given by

$$s(z^1, \ldots, z^n) = (1, z^1, \ldots, z^n),$$

so

$$\omega_{FS} = \sqrt{-1}\partial\bar{\partial} \log(1 + |z^1|^2 + \cdots + |z^n|^2). \tag{1.5}$$

At the origin this equals $\sqrt{-1}\sum_i dz^i \wedge d\bar{z}^i$. The corresponding Hermitian matrix is the identity, which is positive definite.

Example 1.13. If $V \subset \mathbf{CP}^n$ is a projective manifold, then ω_{FS} restricted to V gives a Kähler metric on V since the exterior derivative commutes with pulling back differential forms.

Since the Kähler form ω is a closed real form, it defines a cohomology class $[\omega]$ in $H^2(M, \mathbf{R})$. A fundamental result is the $\partial\bar{\partial}$-lemma, which shows that on a compact manifold, Kähler metrics in a fixed cohomology class can be parametrized by real-valued functions.

Lemma 1.14 ($\partial\bar{\partial}$-lemma). *Let M be a compact Kähler manifold. If ω and η are two real $(1,1)$-forms in the same cohomology class, then there is a function $f : M \to \mathbf{R}$ such that*

$$\eta = \omega + \sqrt{-1}\partial\bar{\partial}f.$$

Proof. The proof of this result requires some ideas from Hodge theory on Kähler manifolds, which we have not discussed. Because of the fundamental nature of the result, we give the proof in any case.

Let g be a Kähler metric on M. Since $[\eta] = [\omega]$ and η, ω are real forms, there exists a real 1-form α such that

$$\eta = \omega + d\alpha.$$

Let us decompose $\alpha = \alpha^{1,0} + \alpha^{0,1}$ into its $(1,0)$ and $(0,1)$ parts, where $\alpha^{0,1} = \overline{\alpha^{1,0}}$ since α is real. Since η, ω are (1,1)-forms, we have

$$\eta = \omega + \bar{\partial}\alpha^{1,0} + \partial\alpha^{0,1} \tag{1.6}$$

and $\partial\alpha^{1,0} = \bar{\partial}\alpha^{0,1} = 0$. The function $\partial^*\alpha^{1,0}$ defined by

$$\partial^*\alpha^{1,0} = -g^{j\bar{k}}\nabla_{\bar{k}}\alpha_j$$

has zero integral on M, so using Theorem 2.12 there is a function f such that

$$\partial^*\alpha^{1,0} = \Delta f = -\partial^*\partial f.$$

Then

$$\partial(\alpha^{1,0} + \partial f) = 0 \quad \text{and} \quad \partial^*(\alpha^{1,0} + \partial f) = 0,$$

so $\alpha^{1,0} + \partial f$ is a ∂-harmonic form. Since g is Kähler, the form is also $\bar{\partial}$-harmonic, so in particular it is $\bar{\partial}$-closed (see Exercise 1.15), so

$$\bar{\partial}\alpha^{1,0} = -\bar{\partial}\partial f.$$

From (1.6) we then have

$$\eta - \omega = -\bar{\partial}\partial f - \partial\overline{\bar{\partial} f} = \partial\bar{\partial}(f - \bar{f}) = 2\sqrt{-1}\partial\bar{\partial}\mathrm{Im}(f),$$

where $\mathrm{Im}(f)$ is the imaginary part of f. □

Exercise 1.15. In the proof of the $\partial\bar{\partial}$-lemma above we used the fact that on a compact Kähler manifold if a $(1,0)$-form α satisfies $\partial\alpha = \partial^*\alpha = 0$, then also $\bar{\partial}\alpha = 0$. Verify this statement by showing that under these assumptions $g^{k\bar{l}}\nabla_k\nabla_{\bar{l}}\alpha_i = 0$ and then integrating by parts. The generalization of this statement is that on a Kähler manifold the ∂- and $\bar{\partial}$-Laplacians coincide (see [**59**, p. 115]).

The next result shows that if we have a Kähler metric, then we can choose particularly nice holomorphic coordinates near any point. This will be very useful in computations later on.

Proposition 1.16 (Normal coordinates). *If g is a Kähler metric, then around any point $p \in M$ we can choose holomorphic coordinates $z^1, \ldots, z^n$ such that the components of g at the point p satisfy*

$$g_{j\bar{k}}(p) = \delta_{jk} \quad \text{and} \quad \frac{\partial}{\partial z^i} g_{j\bar{k}}(p) = \frac{\partial}{\partial \bar{z}^i} g_{j\bar{k}}(p) = 0, \tag{1.7}$$

where δ_{jk} is the identity matrix, i.e. $\delta_{jk} = 0$ if $j \neq k$, and $\delta_{jk} = 1$ if $j = k$.

Proof. It is equivalent to (1.7) to require that the Kähler form satisfies

$$\omega = \sqrt{-1} \sum_{j,k} \left(\delta_{jk} + O(|z|^2)\right) dz^j \wedge d\bar{z}^k, \tag{1.8}$$

where $O(|z|^2)$ denotes terms which are at least quadratic in the $z^i, \bar{z}^i$.

First we choose coordinates w^i such that

$$\omega = \sqrt{-1} \sum_{j,k} \left(\delta_{jk} + \sum_l (a_{j\bar{k}l} w^l + a_{j\bar{k}\bar{l}} \bar{w}^l) + O(|w|^2) \right) dw^j \wedge d\bar{w}^k. \tag{1.9}$$

Next we define new coordinates z^i in a smaller neighborhood of the origin which satisfy

$$w^i = z^i - \frac{1}{2} \sum_{j,k} b_{ijk} z^j z^k$$

for some coefficients b_{ijk} such that $b_{ijk} = b_{ikj}$. Then

$$dw^i = dz^i - \sum_{j,k} b_{ijk} z^j dz^k,$$

so we can compute

$$\omega = \sqrt{-1} \sum_{j,k} \left(\delta_{jk} + \sum_l (a_{j\bar{k}l} z^l + a_{j\bar{k}\bar{l}} \bar{z}^l - b_{klj} z^l - \overline{b_{jlk}} \bar{z}^l) + O(|z|^2) \right) dz^j \wedge d\bar{z}^k.$$

If ω is Kähler, then from (1.9) we know that $a_{j\bar{k}l} = a_{l\bar{k}j}$, so we can define $b_{klj} = a_{j\bar{k}l}$. Then we have

$$a_{j\bar{k}\bar{l}} = \overline{a_{k\bar{j}l}} = \overline{b_{jlk}},$$

so all the linear terms cancel. □

In Riemannian geometry we can always choose normal coordinates in which the first derivatives of the metric vanish at a given point, and of course this result also applies to any Hermitian metric. The point of the previous result is that if the metric is Kähler, then we can even find a *holomorphic* coordinate system in which the first derivatives of the metric vanish at a point. Conversely it is clear from expression (1.8) that if such holomorphic normal coordinates exist, then $d\omega = 0$, so the metric is Kähler.

1.4. Covariant derivatives and curvature

Given a Kähler manifold (M, ω), we use the Levi-Civita connection ∇ to differentiate tensor fields. By definition this satisfies $\nabla g = 0$. In holomorphic normal coordinates the complex structure J is constant, so we obtain $\nabla J = 0$, and since $\omega(X, Y) = g(JX, Y)$, we also have $\nabla \omega = 0$. In terms of local holomorphic coordinates $z^1, \ldots, z^n$, we will use the following notation for the different derivatives:

$$\nabla_i = \nabla_{\partial/\partial z^i}, \quad \nabla_{\bar{i}} = \nabla_{\partial/\partial \bar{z}^i}, \quad \partial_i = \frac{\partial}{\partial z^i}, \quad \partial_{\bar{i}} = \frac{\partial}{\partial \bar{z}^i}.$$

Since

$$J\left(\nabla_j \frac{\partial}{\partial z^k}\right) = \nabla_j J\left(\frac{\partial}{\partial z^k}\right) = \sqrt{-1}\nabla_j \frac{\partial}{\partial z^k},$$

the vector field $\nabla_j \frac{\partial}{\partial z^k}$ has type $(1, 0)$, and so we can define the Christoffel symbols Γ^i_{jk} by

$$\nabla_j \frac{\partial}{\partial z^k} = \sum_i \Gamma^i_{jk} \frac{\partial}{\partial z^i}.$$

For the same reason $\nabla_{\bar{i}} \frac{\partial}{\partial z^k}$ also has type $(1, 0)$, while $\nabla_k \frac{\partial}{\partial \bar{z}^i}$ has type $(0, 1)$. However, since the connection is torsion free,

$$\nabla_{\bar{i}} \frac{\partial}{\partial z^k} = \nabla_k \frac{\partial}{\partial \bar{z}^i},$$

so both vector fields have to vanish. In addition $\nabla_{\bar{i}} T = \overline{\nabla_i \overline{T}}$ for any tensor T, so the connection is determined completely by the coefficients Γ^i_{jk}. Note that

$$\Gamma^i_{jk} = \Gamma^i_{kj} \quad \text{and} \quad \Gamma^{\bar{i}}_{\bar{j}\bar{k}} = \overline{\Gamma^i_{jk}}.$$

Covariant derivatives of tensor fields can be computed using the product rule for derivatives, remembering that on functions the covariant derivatives coincide with the usual partial derivatives.

Example 1.17. To find the covariant derivatives of the form dz^k, we differentiate the relation

$$dz^k\left(\frac{\partial}{\partial z^j}\right) = \delta^k_j,$$

where δ^k_j is the identity matrix. We get

$$(\nabla_i dz^k)\frac{\partial}{\partial z^j} + dz^k\left(\nabla_i \frac{\partial}{\partial z^j}\right) = 0,$$

from which we can calculate that

$$(\nabla_i dz^k)\frac{\partial}{\partial z^j} = -\Gamma^k_{ij},$$

and similarly $(\nabla_i dz^k)\dfrac{\partial}{\partial \bar{z}^j} = 0$. It follows that

$$\nabla_i dz^k = -\sum_j \Gamma^k_{ij} dz^j.$$

From now on we will start using the summation convention, which means that we sum over repeated indices. If we are consistent, then each repeated index should appear once on top and once on the bottom. Usually we will write a tensor such as $a_{i\bar{j}} dz^i \otimes d\bar{z}^j$ (summing over i, j) as just $a_{i\bar{j}}$. Note however that Γ^i_{jk} is not a tensor since it does not transform in the right way under changes of coordinates.

Example 1.18. We compute covariant derivatives of a tensor $a_{i\bar{j}} dz^i \otimes d\bar{z}^j$ using the product rule, namely

$$\begin{aligned}\nabla_{\bar{p}}(a_{i\bar{j}} dz^i \otimes d\bar{z}^j) &= (\partial_{\bar{p}} a_{i\bar{j}}) dz^i \otimes d\bar{z}^j + a_{i\bar{j}} (\nabla_{\bar{p}} dz^i) \otimes d\bar{z}^j + a_{i\bar{j}} dz^i \otimes (\nabla_{\bar{p}} d\bar{z}^j) \\ &= (\partial_{\bar{p}} a_{i\bar{j}}) dz^i \otimes d\bar{z}^j - a_{i\bar{j}} dz^i \otimes (\Gamma^{\bar{j}}_{\bar{p}\bar{l}} d\bar{z}^l) \\ &= \left(\partial_{\bar{p}} a_{i\bar{j}} - \Gamma^{\bar{l}}_{\bar{p}\bar{j}} a_{i\bar{l}}\right) dz^i \otimes d\bar{z}^j.\end{aligned}$$

We can write this formula more concisely as

$$\nabla_{\bar{p}} a_{i\bar{j}} = \partial_{\bar{p}} a_{i\bar{j}} - \Gamma^{\bar{l}}_{\bar{p}\bar{j}} a_{i\bar{l}},$$

and similar formulas for more general tensors can readily be derived.

Lemma 1.19. *In terms of the metric $g_{j\bar{k}}$ the Christoffel symbols are given by*

$$\Gamma^i_{jk} = g^{i\bar{l}} \partial_j g_{k\bar{l}},$$

where $g^{i\bar{l}}$ is the matrix inverse to $g_{i\bar{l}}$.

Proof. The Levi-Civita connection satisfies $\nabla g = 0$. In coordinates this means

$$0 = \nabla_j g_{k\bar{l}} = \partial_j g_{k\bar{l}} - \Gamma^p_{jk} g_{p\bar{l}},$$

so

$$g^{i\bar{l}} \partial_j g_{k\bar{l}} = \Gamma^p_{jk} g_{p\bar{l}} g^{i\bar{l}} = \Gamma^p_{jk} \delta^i_p = \Gamma^i_{jk}. \qquad \square$$

Covariant derivatives do not commute in general, and the failure to commute is measured by the curvature. The curvature is a 4-tensor $R_{i\,k\bar{l}}^{\;j}$, where we will often raise or lower indices using the metric, for example $R_{i\bar{j}k\bar{l}} = g_{p\bar{j}} R_{i\,k\bar{l}}^{\;p}$ (note that the position of the indices is important). The curvature is defined by

$$(\nabla_k \nabla_{\bar{l}} - \nabla_{\bar{l}} \nabla_k) \frac{\partial}{\partial z^i} = R_{i\,k\bar{l}}^{\;j} \frac{\partial}{\partial z^j},$$

while ∇_k commutes with ∇_l and $\nabla_{\bar{k}}$ commutes with $\nabla_{\bar{l}}$.

Exercise 1.20. Verify the following commutation relations for a (0,1)-vector field $v^{\bar{p}}$ and (0,1)-form $\alpha_{\bar{p}}$:

$$(\nabla_k \nabla_{\bar{l}} - \nabla_{\bar{l}} \nabla_k) v^{\bar{p}} = -R^{\bar{p}}_{\ \bar{q}k\bar{l}} v^{\bar{q}},$$
$$(\nabla_k \nabla_{\bar{l}} - \nabla_{\bar{l}} \nabla_k) \alpha_{\bar{p}} = R^{\bar{q}}_{\ \bar{p}k\bar{l}} \alpha_{\bar{q}}.$$

In terms of the Christoffel symbols we can compute

$$R_{i\ k\bar{l}}^{\ j} = -\partial_{\bar{l}} \Gamma^j_{ki},$$

from which we find that in terms of the metric

$$R_{i\bar{j}k\bar{l}} = -\partial_k \partial_{\bar{l}} g_{i\bar{j}} + g^{p\bar{q}} (\partial_k g_{i\bar{q}})(\partial_{\bar{l}} g_{p\bar{j}}).$$

In terms of normal coordinates around a point p we have $R_{i\bar{j}k\bar{l}} = -\partial_k \partial_{\bar{l}} g_{i\bar{j}}$ at p. In other words the curvature tensor of a Kähler metric is the obstruction to finding holomorphic coordinates in which the metric agrees with the Euclidean metric up to second order. It turns out that if we write out the Taylor expansion of the metric in normal coordinates, then each coefficient will only depend on covariant derivatives of the curvature. In particular if the curvature vanishes in a neighborhood of a point, then in normal coordinates the metric is just given by the Euclidean metric.

Exercise 1.21. Verify the following identities for the curvature of a Kähler metric:

$$R_{i\bar{j}k\bar{l}} = R_{i\bar{l}k\bar{j}} = R_{k\bar{j}i\bar{l}} = R_{k\bar{l}i\bar{j}},$$
$$\nabla_p R_{i\bar{j}k\bar{l}} = \nabla_i R_{p\bar{j}k\bar{l}}.$$

Compare these to the identities satisfied by the curvature tensor of a Riemannian metric, in particular the first and second Bianchi identities.

The Ricci curvature is defined to be the contraction

$$R_{i\bar{j}} = g^{k\bar{l}} R_{i\bar{j}k\bar{l}},$$

and the scalar curvature is

$$R = g^{i\bar{j}} R_{i\bar{j}}.$$

Lemma 1.22. *In local coordinates*

$$R_{i\bar{j}} = -\partial_i \partial_{\bar{j}} \log \det(g_{p\bar{q}}).$$

Proof. Using the formulas above, we have

$$\begin{aligned} -\partial_{\bar{j}} \partial_i \log \det(g_{p\bar{q}}) &= -\partial_{\bar{j}} (g^{p\bar{q}} \partial_i g_{p\bar{q}}) \\ &= -\partial_{\bar{j}} \Gamma^p_{ip} \\ &= R_{p\ i\bar{j}}^{\ p} \\ &= R_{i\bar{j}}. \end{aligned}$$

□

As a consequence the Ricci form $\mathrm{Ric}(\omega)$ defined by

$$\mathrm{Ric}(\omega) = \sqrt{-1}R_{i\bar{j}}dz^i \wedge d\bar{z}^j = -\sqrt{-1}\partial\bar{\partial}\log\det(g)$$

in local coordinates is a closed real $(1,1)$-form. Moreover if h is another Kähler metric on M, then $\dfrac{\det(h)}{\det(g)}$ is a globally defined function, so the difference of Ricci forms

$$\mathrm{Ric}(h) - \mathrm{Ric}(g) = -\sqrt{-1}\partial\bar{\partial}\log\frac{\det(h)}{\det(g)}$$

is an exact form. The cohomology class $[\mathrm{Ric}(g)]$ is therefore independent of the choice of Kähler metric. The first Chern class of M is defined to be the cohomology class

$$c_1(M) = \frac{1}{2\pi}[\mathrm{Ric}(g)] \in H^2(M, \mathbf{R}).$$

It turns out that with this normalization $c_1(M)$ is actually an integral cohomology class.

Exercise 1.23. Show that for the Fubini-Study metric ω_{FS} in Example 1.12, the Ricci form satisfies $\mathrm{Ric}(\omega_{FS}) = (n+1)\omega_{FS}$, i.e. that ω_{FS} is a Kähler-Einstein metric.

The fundamental result about the Ricci curvature of Kähler manifolds is Yau's solution of the Calabi conjecture.

Theorem 1.24 (Calabi-Yau theorem)**.** *Let (M, ω) be a compact Kähler manifold, and let α be a real $(1,1)$-form representing $c_1(M)$. Then there exists a unique Kähler metric η on M with $[\eta] = [\omega]$ such that* $\mathrm{Ric}(\eta) = 2\pi\alpha$.

In particular if $c_1(M) = 0$, then every Kähler class contains a unique Ricci flat metric. This provides our first example of a canonical Kähler metric, and it is a very special instance of an extremal metric. We will discuss the proof of this theorem in Section 3.4.

1.5. Vector bundles

A holomorphic vector bundle E over a complex manifold M is a holomorphic family of complex vector spaces parametrized by M. E is itself a complex manifold, together with a holomorphic projection $\pi : E \to M$, and the family is locally trivial so M has an open cover $\{U_\alpha\}$ such that we have biholomorphisms (trivializations)

$$\varphi_\alpha : \pi^{-1}(U_\alpha) \to U_\alpha \times \mathbf{C}^r, \tag{1.10}$$

for some integer $r > 0$ called the rank of E. Under the trivialization φ_α, π corresponds to projection onto U_α. The trivializations are related by holomorphic transition maps

$$\begin{aligned}\varphi_\beta \circ \varphi_\alpha^{-1} : (U_\alpha \cap U_\beta) \times \mathbf{C}^r &\to (U_\alpha \cap U_\beta) \times \mathbf{C}^r \\ (p, v) &\mapsto (p, \varphi_{\beta\alpha}(p)v),\end{aligned} \tag{1.11}$$

which at each point $p \in U_\alpha \cap U_\beta$ gives a linear isomorphism $\varphi_{\beta\alpha}(p)$ from $\mathbf{C}^r$ to $\mathbf{C}^r$. The matrix-valued functions $\varphi_{\beta\alpha}$ satisfy the compatibility (or cocycle) condition

$$\varphi_{\gamma\beta}\varphi_{\beta\alpha} = \varphi_{\gamma\alpha}. \tag{1.12}$$

Conversely any set of holomorphic matrix-valued functions $\varphi_{\beta\alpha}$ satisfying the cocycle conditions defines a vector bundle.

A holomorphic section of a vector bundle E is a holomorphic map $s : M \to E$ such that $\pi \circ s$ is the identity map. A local trivialization φ_α as in (1.10) gives rise to local holomorphic sections corresponding to constant functions on U_α. In particular a basis of $\mathbf{C}^r$ gives rise to local holomorphic sections $\mathbf{s}_1, \ldots, \mathbf{s}_r$ which we call a local holomorphic frame. The values of the $\mathbf{s}_i$ span the fiber $E_p = \pi^{-1}(p)$ at each point $p \in U_\alpha$. All other local holomorphic sections over U_α can be written as

$$f = \sum_{i=1}^r f^i \mathbf{s}_i,$$

where each f^i is a holomorphic function on U_α. We write the space of global holomorphic sections as $H^0(M, E)$ since this forms the first term in a sequence of cohomology spaces $H^i(M, E)$. Although they are fundamental objects, we will not be using these spaces for $i > 0$. An important property which we will discuss later is that $H^0(M, E)$ is finite dimensional if M is compact.

Example 1.25. The $(1, 0)$ part of the cotangent bundle $\Omega^{1,0}M$ is a rank n holomorphic vector bundle over M, where $\dim_{\mathbf{C}} M = n$. In a local chart with holomorphic coordinates $z^1, \ldots, z^n$ a trivialization is given by the holomorphic frame $dz^1, \ldots, dz^n$. The transition map to a different chart is determined by the Jacobian matrix of the coordinate transformation. This bundle is the holomorphic cotangent bundle.

Natural operations on vector spaces can be extended to vector bundles, such as taking tensor products, direct sums, duals, etc.

Example 1.26. On a complex manifold of dimension n we can form the n-th exterior power of the holomorphic cotangent bundle. This is a line bundle (rank 1 vector bundle) denoted by K_M and is called the canonical bundle of M:

$$K_M = \bigwedge^n \Omega^{1,0} M = \Omega^{n,0} M.$$

In local holomorphic coordinates a frame is given by $dz^1 \wedge \cdots \wedge dz^n$, and the transition functions are given by Jacobian determinants.

Exercise 1.27. Show that the canonical line bundle of $\mathbf{CP}^n$ is $K_{\mathbf{CP}^n} = \mathcal{O}(-n-1)$.

Exercise 1.28. Let M be a complex manifold, and suppose that $D \subset M$ is a complex submanifold with (complex) codimension 1. The normal bundle N_D of D in M is defined to be the quotient bundle $(TM|_D)/TD$, where $TM|_D$ is the restriction of the holomorphic tangent bundle of M to D. Show that the canonical bundles of D and M are related by

$$K_D = (K_M|_D) \otimes N_D,$$

where $K_M|_D$ is the restriction of the canonical bundle of M to D. This is called the adjunction formula.

Example 1.29 (Line bundles over $\mathbf{CP}^n$)**.** Since $\mathbf{CP}^n$ is the space of complex lines in $\mathbf{C}^{n+1}$, we can construct a line bundle denoted by $\mathcal{O}(-1)$ over $\mathbf{CP}^n$ by assigning to each point the line it parametrizes. A natural way to think of $\mathcal{O}(-1)$ is as a subbundle of the trivial bundle $\mathbf{CP}^n \times \mathbf{C}^{n+1}$. Recall the charts U_i from Example 1.2. It is a good exercise to work out that under suitable trivializations the transition functions correponding to these charts are given by

$$\varphi_{kj}([Z_0 : \cdots : Z_n]) = \frac{Z_k}{Z_j}, \tag{1.13}$$

in terms of homogeneous coordinates. Note that while Z_j, Z_k are not well-defined functions on $U_j \cap U_k$, their quotient is well-defined.

Since $\mathcal{O}(-1)$ is a subbundle of the trivial bundle, any global holomorphic section of $\mathcal{O}(-1)$ gives rise to a holomorphic map $s : \mathbf{CP}^n \to \mathbf{C}^{n+1}$. The components of s are holomorphic functions on a compact complex manifold, so they are constant. Therefore s itself is a constant map. It is easy to check that non-zero constant maps do not give rise to sections of $\mathcal{O}(-1)$, so $H^0(\mathbf{CP}^n, \mathcal{O}(-1)) = \{0\}$.

The dual of $\mathcal{O}(-1)$ is denoted by $\mathcal{O}(1)$, and by taking tensor powers we obtain line bundles $\mathcal{O}(l)$ for all integers l. The transition functions $\varphi^{(l)}_{kj}$ of $\mathcal{O}(l)$ are given similarly to (1.13) by

$$\varphi^{(l)}_{kj}([Z_0 : \cdots : Z_n]) = \left(\frac{Z_j}{Z_k}\right)^l, \tag{1.14}$$

and the global sections of $\mathcal{O}(l)$ for $l \geqslant 0$ can be thought of as homogeneous polynomials in $Z_0, \ldots, Z_n$ of degree l. In terms of local trivializations, if f is a homogeneous polynomial of degree l, then over the chart U_j we have a holomorphic function $Z_j^{-l} f$. Over different charts these functions patch together using the transition functions (1.14), so they give rise to a global section of $\mathcal{O}(l)$. It turns out that on $\mathbf{CP}^n$ every line bundle is given by $\mathcal{O}(l)$ for some $l \in \mathbf{Z}$.

Exercise 1.30. Suppose that L is a line bundle over a complex manifold M and s is a global holomorphic section of L such that the zero set $s^{-1}(0)$ is a smooth submanifold $D \subset M$. Show that

$$N_D = L|_D,$$

where $L|_D$ is the restriction of L to D and N_D is the normal bundle of D.

Exercise 1.31. Suppose that $M \subset \mathbf{P}^n$ is a smooth hypersurface of degree d, i.e. M is defined by the vanishing of a section of $\mathcal{O}(d)$. Show that if $d > n+1$, then $c_1(M) < 0$, i.e. $-c_1(M)$ is represented by a Kähler metric.

More generally, suppose that $M \subset \mathbf{P}^n$ is a smooth complex submanifold of codimension r, defined by the intersection of r hypersurfaces of degrees $d_1, \ldots, d_r$. If $d_1 + \cdots + d_r > n+1$, then show that $c_1(M) < 0$.

1.6. Connections and curvature of line bundles

The Levi-Civita connection that we used before is a canonical connection on the tangent bundle of a Riemannian manifold. Analogously there is a canonical connection on an arbitrary holomorphic vector bundle equipped with a Hermitian metric, called the Chern connection.

A Hermitian metric h on a complex vector bundle is a smooth family of Hermitian inner products on the fibers. In other words for any two local sections (not necessarily holomorphic) s_1, s_2, we obtain a function $\langle s_1, s_2 \rangle_h$ which satisfies $\langle s_2, s_1 \rangle_h = \overline{\langle s_1, s_2 \rangle_h}$. We can think of the inner product as a section of $E^* \otimes \overline{E^*}$. The Chern connection on a holomorphic vector bundle is then the unique connection on E such that the derivative of the inner product is zero and $\nabla_{\bar{i}} s = 0$ for any local holomorphic section of E. The derivative of the inner product h is zero if and only if

$$\partial_k(\langle s_1, s_2 \rangle_h) = \langle \nabla_k s_1, s_2 \rangle_h + \langle s_1, \nabla_{\bar{k}} s_2 \rangle_h,$$

and a similar formula holds for $\partial_{\bar{k}}$. Just as before, covariant derivatives do not commute in general, and the curvature $F_{k\bar{l}}$ is defined by

$$F_{k\bar{l}} = \nabla_k \nabla_{\bar{l}} - \nabla_{\bar{l}} \nabla_k.$$

The $F_{k\bar{l}}$ are the components of an endomorphism-valued $(1,1)$-form.

Exercise 1.32. Show that if (E, h_E) and (F, h_F) are Hermitian holomorphic bundles whose curvature forms are R_E and R_F, respectively, then the curvature of the tensor product $(E \otimes F, h_E \otimes h_F)$ is the sum

$$R_E \otimes \mathrm{Id}_F + \mathrm{Id}_E \otimes R_F,$$

where Id_E and Id_F are the identity endomorphisms of E, F.

Note that on a complex manifold M with a Hermitian metric, the holomorphic tangent bundle $T^{1,0}M$ has two natural connections. Identifying $T^{1,0}M$ with the real tangent bundle TM, there is the Levi-Civita connection that we were concerned with in Section 1.4, and there is also the Chern connection.

Exercise 1.33. Show that a Hermitian metric is Kähler if and only if the Levi-Civita and Chern connections coincide on $T^{1,0}M$. In this case the curvature tensor $R_{i\ k\bar{l}}^{\ j}$ defined earlier is the same as the curvature $F_{k\bar{l}}$ of the Chern connection on $T^{1,0}$ except that in the latter the endomorphism part is suppressed in the notation.

Let us focus now on the case of line bundles since in this book we will mainly be concerned with those. On a line bundle a Hermitian metric at any point is determined by the norm of any given non-vanishing section at that point. Let s be a local non-vanishing holomorphic section of L, and write

$$h(s) = \langle s, s \rangle_h.$$

Then locally any other section of L can be written as fs for some function f, and the norm of fs is $|fs|_h^2 = |f|^2 h(s)$. In particular we have functions A_k (analogous to the earlier Christoffel symbols) defined by

$$\nabla_k s = A_k s.$$

Then the curvature is determined by (remembering that s is holomorphic)

$$F_{k\bar{l}} s = -\nabla_{\bar{l}} \nabla_k s = -\nabla_{\bar{l}}(A_k s) = -(\partial_{\bar{l}} A_k) s,$$

so $F_{k\bar{l}} = -\partial_{\bar{l}} A_k$. To determine A_k we use the defining properties of the Chern connection to get

$$\partial_k h(s) = \langle \nabla_k s, s \rangle_h = A_k h(s),$$

so $A_k = h(s)^{-1} \partial_k h(s)$. It follows that

$$F_{k\bar{l}} = -\partial_{\bar{l}}(h(s)^{-1} \partial_k h(s)) = -\partial_{\bar{l}} \partial_k \log h(s).$$

We can summarize these calculations as follows.

Lemma 1.34. *The curvature of the Chern connection of a holomorphic line bundle equipped with a Hermitian metric is given by*

$$F_{k\bar{l}} = -\partial_k \partial_{\bar{l}} \log h(s),$$

where $h(s) = \langle s, s\rangle_h$ for a local holomorphic section s.

Note the similarity with Lemma 1.22 dealing with the Ricci curvature. The relationship between the two results is that as we remarked above, the Levi-Civita connection of a Kähler metric coincides with the Chern connection on its holomorphic tangent bundle. The determinant of the metric defines a Hermitian metric on the top exterior power $\bigwedge T^{1,0}$, and the Ricci curvature is the curvature of the induced connection on this line bundle.

Just as in the case of the Ricci curvature, Lemma 1.34 implies that the form locally defined by

$$F(h) = \sqrt{-1}F_{k\bar{l}}dz^k \wedge d\bar{z}^l = -\sqrt{-1}\partial\bar{\partial} \log h(s)$$

is a closed real $(1,1)$-form. Any other Hermitian metric can be written as $e^{-f}h$ for a globally defined function f, and we can check that

$$F(e^{-f}h) - F(h) = \sqrt{-1}\partial\bar{\partial} f, \tag{1.15}$$

so if we choose a different Hermitian metric on L, then $F(h)$ changes by an exact form. This allows us to define the first Chern class of the line bundle L to be

$$c_1(L) = \frac{1}{2\pi}[F(h)] \in H^2(M, \mathbf{R}).$$

The $\partial\bar{\partial}$-lemma and (1.15) imply that every real (1,1)-form in $c_1(L)$ is the curvature of some Hermitian metric on L.

Remark 1.35. The normalizing factor of 2π is chosen because it turns out that this way $c_1(L)$ is an integral cohomology class. We will not need this, but it is an important fact about characteristic classes. See [**59**, p. 139].

For us the most important property that a line bundle can have is the positivity of its curvature.

Definition 1.36. Let us call a real (1,1)-form positive if the symmetric bilinear form $(X, Y) \mapsto \alpha(X, JY)$ defined for real tangent vectors X, Y is positive definite. For instance the Kähler form of a Kähler metric is positive.

A cohomology class in $H^2(M, \mathbf{R})$ is called positive if it can be represented by a closed positive (1,1)-form. Finally we call a line bundle positive if its first Chern class is positive. Equivalently a line bundle is positive if for a suitable Hermitian metric h the curvature form $F(h)$ is a Kähler form.

Example 1.37. The $\mathcal{O}(-1)$ bundle over $\mathbf{CP}^n$ has a natural Hermitian metric h since it is a subbundle of the trivial bundle $\mathbf{CP}^n \times \mathbf{C}^{n+1}$ on which we can use the standard Hermitian metric of $\mathbf{C}^{n+1}$. On the open set U_0 if we use coordinates $z^i = \dfrac{Z_i}{Z_0}$ for $i = 1, \ldots, n$, then a holomorphic section of $\mathcal{O}(-1)$ over U_0 is given by

$$s : (z^1, \ldots, z^n) \mapsto (1, z^1, \ldots, z^n) \in \mathbf{C}^{n+1}$$

since $(z^1, \ldots, z^n) \in U_0$ corresponds to the point $[1 : z^1 : \cdots : z^n]$ in homogeneous coordinates. By Lemma 1.34 the curvature form of h is then

$$F(h) = -\sqrt{-1}\partial\overline{\partial} \log h(s) = -\sqrt{-1}\partial\overline{\partial} \log(1 + |z^1|^2 + \cdots + |z^n|^2),$$

so $F(h) = -\omega_{FS}$ in terms of the Fubini-Study metric of Example 1.12.

The metric h induces a metric on the dual bundle $\mathcal{O}(1)$, whose curvature form will then be ω_{FS}. Since this is a Kähler form, $\mathcal{O}(1)$ is a positive line bundle.

In this book, just as we will restrict our attention to compact complex manifolds which are submanifolds of projective space, we will generally also restrict our attention to Kähler metrics whose Kähler class is the first Chern class of a line bundle. The Kodaira embedding theorem in the next section states that if a compact complex manifold admits such a Kähler metric, then it is automatically a projective manifold.

1.7. Line bundles and projective embeddings

Suppose that $L \to M$ is a holomorphic line bundle over a complex manifold M. If $s_0, \ldots, s_k$ are sections of L, then over the set $U \subset M$ where at least one s_i is non-zero, we obtain a holomorphic map

$$\begin{aligned} U &\to \mathbf{CP}^k \\ p &\mapsto [s_0(p) : \cdots : s_k(p)]. \end{aligned} \tag{1.16}$$

Definition 1.38. A line bundle L over M is very ample if for suitable sections $s_0, \ldots, s_k$ of L the map (1.16) defines an embedding of M into $\mathbf{CP}^k$. A line bundle L is ample if for a suitable integer $r > 0$ the tensor power L^r is very ample.

Example 1.39. The bundle $\mathcal{O}(1)$ over $\mathbf{CP}^n$ is very ample, and the sections $Z_0, \ldots, Z_n$ from Example 1.29 define the identity map from $\mathbf{CP}^n$ to itself. More generally for any projective manifold $V \subset \mathbf{CP}^n$, the restriction of $\mathcal{O}(1)$ to V is a very ample line bundle. Conversely if L is a very ample line bundle over V, then L is isomorphic to the restriction of the $\mathcal{O}(1)$ bundle under a projective embedding furnished by sections of L.

The following is a fundamental result relating the curvature of a line bundle to ampleness.

Theorem 1.40 (Kodaira embedding theorem). *Let L be a line bundle over a compact complex manifold M. Then L is ample if and only if the first Chern class $c_1(L)$ is positive.*

The difficult implication is that a line bundle with positive first Chern class is ample. The proof requires showing that a sufficiently high power of the line bundle admits enough holomorphic sections to give rise to an embedding of the manifold, but it is already non-trivial to show that there is at least one non-zero holomorphic section. One way to proceed is through Kodaira's vanishing theorem for cohomology (see [**59**, p. 189]). Another approach is through studying the Bergman kernel for large powers of the line bundle L, which we will discuss in Chapter 7. See in particular Exercise 7.10.

Example 1.41. Let L be the trivial line bundle over $\mathbf{C}^n$, so holomorphic sections of L are simply holomorphic functions on $\mathbf{C}^n$. Write $\mathbf{1}$ for the section given by the constant function 1. For $k > 0$ let us define the Hermitian metric h so that $h(\mathbf{1}) = e^{-k|z|^2}$. Then by Lemma 1.34,

$$F(h) = k\sqrt{-1}\sum_{i=1}^{n} dz^i \wedge d\bar{z}^i.$$

When k is very large, then on the one hand the section $\mathbf{1}$ decays very rapidly as we move away from the origin, and on the other hand the curvature of the line bundle is very large. The idea of Tian's argument [**110**] which we will explain in Section 7.2 is that if the curvature of a line bundle L at a point p is very large, then using a suitable cutoff function, we can glue the rapidly decaying holomorphic section $\mathbf{1}$ into a neighborhood around p. Because of the cutoff function this will no longer be holomorphic, but the error is sufficiently small so that it can be corrected to obtain a holomorphic section of L, which is "peaked" at p. If the curvature of the line bundle is large everywhere, then this construction will give rise to enough holomorphic sections to embed the manifold into projective space.

A much simpler result is that if the line bundle L over a compact Kähler manifold is negative, i.e. $-c_1(L)$ is a positive class, then there are no non-zero holomorphic sections at all. To see this, choose a Hermitian metric on L whose curvature form $F_{k\bar{l}}$ is negative definite. From the definition of the curvature, the Chern connection satisfies

$$\nabla_k\nabla_{\bar{l}} = \nabla_{\bar{l}}\nabla_k + F_{k\bar{l}}.$$

If s is a global holomorphic section of L over M, then we have

$$0 = \langle g^{k\bar{l}}\nabla_k\nabla_{\bar{l}}s, s\rangle_h = \langle g^{k\bar{l}}\nabla_{\bar{l}}\nabla_k s, s\rangle_h + g^{k\bar{l}}F_{k\bar{l}}|s|_h^2 \leqslant \langle g^{k\bar{l}}\nabla_{\bar{l}}\nabla_k s, s\rangle_h - c|s|_h^2$$

for some constant $c > 0$ since $F_{k\bar{l}}$ is negative definite. Integrating this over M and integrating by parts we get

$$(1.17) \qquad 0 \leqslant -\int_M |\nabla s|^2_{g\otimes h}\, dV - c\int_M |s|^2_h\, dV.$$

Here dV is the volume form of the metric g, and we are writing $g\otimes h$ for the natural Hermitian metric on $T^{1,0}\otimes L$, which in coordinates can be written as

$$|\nabla s|^2_{g\otimes h} = g^{k\bar{l}} h \nabla_k s \overline{\nabla_l s}.$$

From the inequality (1.17) it is clear that we must have $s = 0$.

It is perhaps instructive to work out the integration by parts carefully to familiarize oneself with the notation. Note first of all that using the Levi-Civita connection together with the Chern connection of L, we obtain natural connections on any vector bundle related to $T^{1,0}M$ and L, and their tensor products, direct sums, etc. Now let us define the vector field $v^{\bar{l}}$ by

$$v^{\bar{l}} = g^{k\bar{l}} h(\nabla_k s)\overline{s}.$$

Note that $g^{j\bar{k}}$ is a section of $T^{1,0}M \otimes T^{0,1}M$, h is a section of $L^*\otimes\overline{L^*}$, $\nabla_k s$ is a section of $\Omega^{1,0}M\otimes L$, and $\overline{s}$ is a section of $\overline{L}$. The section $v^{\bar{l}}$ of $T^{0,1}M$ is obtained by taking the tensor product of these four sections and performing various contractions between pairwise dual spaces. The function $\nabla_{\bar{l}} v^{\bar{l}}$ is the divergence of a vector field, so it has integral zero (it is the exterior derivative $d(\iota_v dV)$ of the contraction of the volume form with v, so we can use Stokes's theorem). Using the product rule, we have

$$\nabla_{\bar{l}} v^{\bar{l}} = (\nabla_{\bar{l}} g^{k\bar{l}}) h(\nabla_k s)\overline{s} + g^{k\bar{l}}(\nabla_{\bar{l}} h)(\nabla_k s)\overline{s} + g^{k\bar{l}} h(\nabla_{\bar{l}}\nabla_k s)\overline{s} + g^{k\bar{l}} h(\nabla_k s)(\nabla_{\bar{l}}\overline{s}),$$

where each time the covariant derivative of the appropriate bundle is used. By the defining properties of the Levi-Civita and Chern connections, we have $\nabla g = 0$ and $\nabla h = 0$, so

$$\nabla_{\bar{l}} v^{\bar{l}} = g^{k\bar{l}} h(\nabla_{\bar{l}}\nabla_k s)\overline{s} + g^{k\bar{l}} h(\nabla_k s)(\nabla_{\bar{l}}\overline{s})$$

Integrating this equation gives the integration by parts formula:

$$\int_M g^{k\bar{l}} h(\nabla_{\bar{l}}\nabla_k s)\overline{s}\, dV = -\int_M g^{k\bar{l}} h(\nabla_k s)(\nabla_{\bar{l}}\overline{s})\, dV.$$

Exercise 1.42. Let L be a holomorphic line bundle on a connected compact Kähler manifold with $c_1(L) = 0$. Show that if L is not the trivial line bundle, then it has no non-zero global holomorphic sections.

Exercise 1.43. A holomorphic vector field is a section v^i of $T^{1,0}M$ such that $\nabla_{\bar{k}} v^i = 0$. Show that on a compact Kähler manifold with negative definite Ricci form there are no non-zero holomorphic vector fields.

Chapter 2

Analytic Preliminaries

In this section we collect some fundamental results about elliptic operators on manifolds, which we will need later on. The most important results for us will be the Schauder estimates theorem (Theorem 2.10) and the solution of linear elliptic equations on compact manifolds (Theorem 2.13). The basic reference for elliptic equations of second order is Gilbarg-Trudinger [**58**]. For analysis on manifolds Aubin [**10**] gives an overview and Donaldson [**41**] is also a good resource.

2.1. Harmonic functions on $\mathbf{R}^n$

Let $U \subset \mathbf{R}^n$ be an open set. A function $f : U \to \mathbf{R}$ is called harmonic if

$$\Delta f := \frac{\partial^2 f}{\partial x^1 \partial x^1} + \cdots + \frac{\partial^2 f}{\partial x^n \partial x^n} = 0 \quad \text{on } U.$$

For any $x \in \mathbf{R}^n$ let us write $B_r(x)$ for the open r-ball around x. For short we will write $B_r = B_r(0)$. The most basic property of harmonic functions is the following.

Theorem 2.1 (Mean value theorem). *If $f : U \to \mathbf{R}$ is harmonic, $x \in U$, and the r-ball $B_r(x) \subset U$, then*

$$f(x) = \frac{1}{\mathrm{Vol}(\partial B_r)} \int_{\partial B_r(x)} f(y)\, dy.$$

Proof. For $\rho \leqslant r$ let us define

$$F(\rho) = \int_{\partial B_1} f(x + \rho y)\, dy.$$

Then

$$\begin{aligned} F'(\rho) &= \int_{\partial B_1} \nabla f(x+\rho y)\cdot y\, dy \\ &= \int_{B_1} \Delta f(x+\rho y)\, dy = 0, \end{aligned}$$

where we used Green's theorem. This means that F is constant, but also by changing variables

$$F(r) = \frac{\mathrm{Vol}(\partial B_1)}{\mathrm{Vol}(\partial B_r)} \int_{\partial B_r(x)} f(y)\, dy,$$

while $\lim_{\rho\to 0} F(\rho) = \mathrm{Vol}(\partial B_1) f(x)$. □

By averaging the mean value property over spheres of different radii, we obtain the following.

Corollary 2.2. *Let $\eta : \mathbf{R}^n \to \mathbf{R}$ be smooth, radially symmetric, supported in B_1, and let $\int \eta = 1$. If $f : B_2 \to \mathbf{R}$ is harmonic, then for all $x \in B_1$ we have*

$$f(x) = \int_{\mathbf{R}^n} f(x-y)\eta(y)\, dy = \int_{\mathbf{R}^n} f(y)\eta(x-y)\, dy.$$

An important consequence is that the L^1-norm of a harmonic function on B_2 controls all the derivatives of the function on the smaller ball B_1.

Corollary 2.3. *There are constants C_k such that if $f : B_2 \to \mathbf{R}$ is harmonic, then*

$$\sup_{B_1} |\nabla^k f| \leqslant C_k \int_{B_2} |f(y)|\, dy.$$

In particular even if f is only assumed to be twice differentiable, it follows that f is smooth on B_1.

Proof. For any $x \in B_1$ we can use the previous corollary and differentiate under the integral sign to get

$$\nabla^k f(x) = \int_{\mathbf{R}^n} f(y)\nabla^k \eta(x-y)\, dy,$$

so

$$|\nabla^k f(x)| \leqslant (\sup_{B_1} |\nabla^k \eta|) \int_{B_2} |f(y)|\, dy.$$

The result follows with $C_k = \sup |\nabla^k \eta|$. □

This interior regularity result together with a scaling argument implies the following "rigidity" statement. In the next section we will see that conversely this rigidity statement can be used to derive interior regularity results.

Corollary 2.4 (Liouville's theorem). *We say that a function $f : \mathbf{R}^n \to \mathbf{R}$ has sublinear growth if*

$$\lim_{R\to\infty} R^{-1} \sup_{B_R} |f| = 0.$$

If f is harmonic on $\mathbf{R}^n$ and has sublinear growth, then f is constant.

Proof. For $r > 0$ let $f_r(x) = f(rx)$, which is also harmonic. The previous corollary implies that

$$|\nabla f_r(0)| \leqslant C_1 \int_{B_2} |f_r(x)|\, dx \leqslant C' \sup_{B_2} |f_r| = C' \sup_{B_{2r}} |f|,$$

for some constant C'. But $\nabla f_r(0) = r\nabla f(0)$, so we get

$$|\nabla f(0)| \leqslant C' r^{-1} \sup_{B_{2r}} |f|$$

for all $r > 0$. Taking $r \to \infty$, this implies $\nabla f(0) = 0$. By translating f, we get $\nabla f(x) = 0$ for all x, so f is constant. □

An alternative approach to the proof of this result is to expand the function f in terms of spherical harmonics, as we will do in the proof of Theorem 8.3. Indeed that argument shows that if $f : \mathbf{R}^n \to \mathbf{R}$ is harmonic, then we can write $f = c + f_{2-n}$, where c is a constant and $f_{2-n}(x) = O(|x|^{2-n})$. If $n > 1$, we can then integrate by parts to get

$$0 = \int_{\mathbf{R}^n} f_{2-n}(x) \Delta f_{2-n}(x), \quad dx = -\int_{\mathbf{R}^n} |\nabla f_{2-n}|^2(x)\, dx,$$

from which it follows that $f_{2-n} = 0$. The advantage of this approach is that it also applies to higher-order operators, such as Δ^2, which can be used to obtain Schauder estimates for fourth-order elliptic operators using our method of proof in Section 2.3 below.

Exercise 2.5. Let $f : \mathbf{R}^n \to \mathbf{R}$ be a harmonic function such that for some constant C we have $|f(x)| \leqslant C(1+|x|)^k$ for all x. Show that f is a polynomial of degree at most k.

2.2. Elliptic differential operators

In Riemannian geometry many of the natural differential equations that arise are elliptic. We will focus on scalar equations of second order. A general linear differential operator of second order is of the form

$$L(f) = \sum_{j,k=1}^{n} a_{jk} \frac{\partial^2 f}{\partial x^j \partial x^k} + \sum_{l=1}^{n} b_l \frac{\partial f}{\partial x^l} + cf, \tag{2.1}$$

where $f, a_{jk}, b_l, c : \Omega \to \mathbf{R}$ are all functions on an open set $\Omega \subset \mathbf{R}^n$. This operator is elliptic if the matrix (a_{jk}) is positive definite.

From now on we will assume that the a_{jk}, b_l, c are all smooth functions. In addition we will usually work on a compact manifold (in which Ω is a coordinate chart), so we will be able to assume the stronger condition of uniform ellipticity:

$$\lambda|v|^2 \leqslant \sum_{j,k=1}^{n} a_{jk}(x)v^j v^k \leqslant \Lambda|v|^2 \quad \text{for all } x \in \Omega$$

for all vectors v and some constants $\lambda, \Lambda > 0$.

While we assume the coefficients of our operator to be smooth, in constructing solutions to linear equations it is usually easiest to first obtain a weak solution. Weak solutions are defined in terms of the formal adjoint L^* of L, which is the operator

$$L^*(f) = \sum_{j,k=1}^{n} \frac{\partial^2}{\partial x^j \partial x^k}(a_{jk}f) - \sum_{l=1}^{n} \frac{\partial}{\partial x^l}(b_l f) + cf.$$

We say that a function f which is locally integrable on Ω is a weak solution of the equation $L(f) = g$ if

$$\int_\Omega f L^*(\varphi)\, dV = \int_\Omega g\varphi\, dV$$

for all compactly supported smooth functions φ on Ω, where dV is the usual volume measure on $\mathbf{R}^n$. The adjoint is defined so that if f is a weak solution of $L(f) = g$ and f is actually smooth, then integration by parts shows that $L(f) = g$ in the usual sense. A fundamental property of elliptic operators is that weak solutions are automatically smooth.

Theorem 2.6. *Suppose that f is a weak solution of the equation $L(f) = g$, where L is a linear elliptic operator with smooth coefficients and g is a smooth function. Then f is also smooth.*

There are many more general regularity statements, but for us this simple one will suffice. The proof is somewhat involved and requires techniques that we will not use in the rest of the book. One approach to the proof is to first use convolutions to construct smoothings f_ε of f and then derive estimates for the f_ε in various Sobolev spaces which are independent of ε. The Sobolev embedding theorem will then ensure that f, which is the limit of the f_ε as $\varepsilon \to 0$, is smooth. For details of this approach, see for example Griffiths-Harris [**59**, p. 380].

2.3. Schauder estimates

In Section 2.2 we saw that solutions of elliptic equations have very strong regularity properties. In this section we will see a more refined version of this idea. We will once again work in a domain $\Omega \subset \mathbf{R}^n$. More precisely we should generally work on a bounded open set with at least C^1-boundary, but not much is lost by assuming that Ω is simply an open ball in $\mathbf{R}^n$. Recall that for $\alpha \in (0,1)$ the C^α Hölder coefficient of a function f on Ω is defined as

$$|f|_{C^\alpha} = \sup_{x,y\in\Omega, x\neq y} \frac{|f(x)-f(y)|}{|x-y|^\alpha}.$$

Using this we can define the $C^{k,\alpha}$-norms for $k \in \mathbf{N}$ and $\alpha \in (0,1)$ as

$$\|f\|_{C^{k,\alpha}} = \sup_{|\mathbf{l}|\leqslant k, x\in\Omega} |\partial^{\mathbf{l}} f(x)| + \sup_{|\mathbf{l}|=k} |\partial^{\mathbf{l}} f|_{C^\alpha},$$

where $\mathbf{l} = (l_1, \ldots, l_n)$ is a multi-index and

$$\partial^{\mathbf{l}} = \frac{\partial}{\partial x^{l_1}} \cdots \frac{\partial}{\partial x^{l_n}}$$

is the corresponding partial derivative of order $|\mathbf{l}| = l_1 + \cdots + l_n$. The space $C^{k,\alpha}(\Omega)$ is the space of functions on Ω whose $C^{k,\alpha}$-norm is finite. If $k > 0$, then such functions are necessarily k-times continuously differentiable. Moreover $C^{k,\alpha}(\Omega)$ is complete, i.e. any Cauchy sequence with respect to the $C^{k,\alpha}$ norm converges in $C^{k,\alpha}$.

Of crucial importance is the following consequence of the Arzela-Ascoli theorem.

Theorem 2.7. *Suppose that Ω is a bounded set and $u_k : \Omega \to \mathbf{R}$ is a sequence of functions such that $\|u_k\|_{C^{k,\alpha}} < C$ for some constant C. Then a subsequence of the u_k is convergent in $C^{l,\beta}$ for any l, β such that $l+\beta < k+\alpha$.*

Let us suppose again that

$$L(f) - \sum_{j,k=1}^n a_{jk} \frac{\partial^2 f}{\partial x^j \partial x^k} + \sum_{l=1}^n b_l \frac{\partial f}{\partial x^l} + cf$$

is a uniformly elliptic second-order differential operator with smooth coefficients. In particular we have the inequalities

$$\lambda|v|^2 \leqslant \sum_{j,k-1}^n a_{jk}(x) v^j v^k \leqslant \Lambda |v|^2 \quad \text{for all } x \in \Omega \tag{2.2}$$

for some $\lambda, \Lambda > 0$.

Theorem 2.8 (Local Schauder estimates). *Let Ω be a bounded domain, let $\Omega' \subset \Omega$ be a smaller domain with the distance $d(\Omega', \partial\Omega) > 0$, and suppose that $\alpha \in (0,1)$ and $k \in \mathbf{N}$. There is a constant C such that if $L(f) = g$, then we have*

$$\|f\|_{C^{k+2,\alpha}(\Omega')} \leqslant C(\|g\|_{C^{k,\alpha}(\Omega)} + \|f\|_{C^0(\Omega)}). \tag{2.3}$$

Moreover C only depends on k,α, the domains Ω,Ω', the $C^{k,\alpha}$-norms of the coefficients of L, and the constants of ellipticity λ,Λ in (2.2).

Sketch of proof. There are several approaches to the proof, usually reducing the problem to the case when the coefficients of L are constant (see Gilbarg-Trudinger [**58**]). We will use an argument by contradiction, using Liouville's theorem for harmonic functions (see Simon [**96**]). This type of "blow-up" argument is very common in geometric analysis.

We will only treat the case $k=0$ since the general case can be reduced to this by differentiating the equation k times. Moreover we will only treat operators L with $b_l, c=0$ since again the general case can be reduced to this one (see [**58**, Section 6.8]).

First we show a weaker estimate, namely that under the assumptions of the theorem, we have a constant C such that

$$\|f\|_{C^{2,\alpha}(\Omega')} \leqslant C(\|g\|_{C^\alpha(\Omega)} + \|f\|_{C^2(\Omega)}). \tag{2.4}$$

More precisely for any $x\in\Omega$ we let $d_x = \min\{1, d(x,\partial\Omega)\}$. We will show that for some constant C we have

$$\min\{d_x,d_y\}^\alpha \frac{|\partial^{\mathbf{l}} f(x) - \partial^{\mathbf{l}} f(y)|}{|x-y|^\alpha} \leqslant C(\|g\|_{C^\alpha(\Omega)} + \|f\|_{C^2(\Omega)}), \tag{2.5}$$

for all $x,y\in\Omega$ and $\partial^{\mathbf{l}}$ any second-order partial derivative.

To argue by contradiction let us fix constants K,λ,Λ and suppose that for arbitrary C there exist functions a_{jk}, f, g on Ω satisfying the equation

$$\sum_{j,k} a_{jk}\frac{\partial^2 f}{\partial x^j \partial x^k} = g$$

such that in addition the a_{jk} satisfy $\|a_{jk}\|_{C^\alpha} \leqslant K$, the uniform ellipticity condition (2.2) holds, and $\|g\|_{C^\alpha}, \|f\|_{C^2} \leqslant 1$. Moreover there are points $p,q\in\Omega$ and a second-order partial derivative $\partial^{\mathbf{l}}$ such that

$$\min\{d_p,d_q\}^\alpha \frac{|\partial^{\mathbf{l}} f(p) - \partial^{\mathbf{l}} f(q)|}{|p-q|^\alpha} = C, \tag{2.6}$$

and at the same time C is the largest possible value for this expression for other choices of the points and $\mathbf{l}$. For these points let us write

$$\frac{|\partial^{\mathbf{l}} f(p) - \partial^{\mathbf{l}} f(q)|}{|p-q|^\alpha} = M \geqslant C$$

and $r=|p-q|$. We define the rescaled function $\tilde{f}(x) = M^{-1}r^{-2-\alpha}f(p+rx)$ and let

$$F(x) = \tilde{f}(x) - \tilde{f}(0) - \sum_k x^k \partial_k \tilde{f}(0) - \frac{1}{2}\sum_{j,k} x^j x^k \partial_j\partial_k \tilde{f}(0).$$

Then this function F satisfies the following properties:

(i) F is defined on (at least) a ball of radius d_p/r around the origin.

(ii) $F(0) = \partial F(0) = \partial^2 F(0) = 0$, where ∂^2 means any second-order derivative.

(iii) On the ball of radius $d_p/(2r)$ around the origin, we have $|\partial^2 F|_{C^\alpha} \leqslant 2^\alpha$.

(iv) For $y = r^{-1}(q-p)$ we have $|y| = 1$ and

$$|\partial^{\mathbf{l}} F(0) - \partial^{\mathbf{l}} F(y)| = 1.$$

(v) F satisfies the equation

$$\sum_{j,k} a_{jk}(p+rx)\frac{\partial^2 F(x)}{\partial x^j \partial x^k} = M^{-1}r^{-\alpha}(g(p+rx) - g(p))$$
$$+M^{-1}r^{-\alpha}\sum_{j,k}(a_{jk}(p) - a_{jk}(p+rx))\frac{\partial^2 f(p)}{\partial x^j \partial x^k}.$$

Now suppose that we can perform this construction for larger and larger C, obtaining a sequence of functions $F^{(i)}$ as above, together with $a_{jk}^{(i)}$, $g^{(i)}$, unit vectors $y^{(i)}$, and second-order partial derivatives $\partial^{\mathbf{l}_i}$. After choosing a subsequence we can assume that the $y^{(i)}$ converge to a vector y and the second-order partial derivatives are all the same $\partial^{\mathbf{l}}$. Since we have assumed that $\|f\|_{C^2} \leqslant 1$, from (2.6) we see that $rd_p^{-1} \to 0$ as $C \to \infty$, so the $F^{(i)}$ are defined on larger and larger balls. From properties (ii) and (iii) the $F^{(i)}$ satisfy uniform $C^{2,\alpha}$-bounds on fixed balls, so on each fixed ball we can extract a convergent subsequence in C^2. By a diagonal argument we obtain a function $G : \mathbf{R}^n \to \mathbf{R}$ which on each fixed ball is a C^2-limit of a subsequence of the $F^{(i)}$, and in particular G satisfies the conditions (i)–(iv).

The "stretched" functions $x \mapsto a_{jk}^{(i)}(p+rx)$ satisfy uniform C^α-bounds, so by choosing a further subsequence we can assume that they converge to functions A_{jk}, which because of the stretching are actually constant. The uniform C^α-bounds on the $g^{(i)}$ and $a_{jk}^{(i)}$, the assumption that $\|f\|_{C^2} \leqslant 1$, together with property (v) imply that

$$\sum_{j,k} A_{jk}\frac{\partial^2 G}{\partial x^j \partial x^k} = 0.$$

After a linear change of coordinates by a matrix T, we obtain a harmonic function $H(x) = G(Tx)$ defined on all of $\mathbf{R}^n$. By Corollary 2.3 or Theorem 2.6 the function H is smooth, and so its second derivatives $\partial^{\mathbf{l}} H$ are also harmonic. From properties (ii) and (iii) we have $|\partial^{\mathbf{l}} G(x)| \leqslant 2^\alpha |x|^\alpha$, so each second derivative of H is a harmonic function with sublinear growth, which is therefore constant by Corollary 2.4. This implies that $\partial^{\mathbf{l}}G$ is identically zero, contradicting property (iv). This proves the estimate (2.4).

From (2.5) one can deduce the estimate we need (replacing the C^2-norm of f by the C^0-norm) by using another argument by contradiction as follows. Still under the same assumptions as in the statement of the theorem, we will now show that there is a constant C such that

$$d_x^2|\partial^{\mathbf{l}}f(x)| \leqslant C(\|g\|_{C^\alpha(\Omega)} + \|f\|_{C^0(\Omega)}),$$

for all $x \in \Omega$ and second-order derivative $\partial^{\mathbf{l}}$. We use a very similar argument to before. Suppose that $\|g\|_{C^\alpha}, \|f\|_{C^0} \leqslant 1$. Choose $p \in \Omega$ and a multi-index $\mathbf{l}$ such that

$$d_p^2|\partial^{\mathbf{l}}f(p)| = C$$

and C is the largest possible value of this expression. Let

$$|\partial^{\mathbf{l}}f(p)| = M \geqslant C.$$

Define the rescaled function $F(x) = M^{-1}d_p^{-2}f(p + d_p x)$. Then F satisfies the following:

(i) F is defined at least on a ball of radius 1 around the origin.

(ii) On the ball of radius 1/2 around the origin we have $\|F\|_{C^2} \leqslant K$ for some fixed constant K.

(iii) $|\partial^{\mathbf{l}}F(0)| = 1$.

(iv) F satisfies the equation

$$\sum_{j,k} a_{jk}(p + d_p x)\frac{\partial^2 F(x)}{\partial x^j \partial x^k} = M^{-1}g(p + d_p x).$$

(v) $|F| \leqslant M^{-1}d_p^{-2}\|f\|_{C^0}$.

If we have a family of such functions $F^{(i)}$ with larger and larger C, then since $d_p \leqslant 1$ and $d_p^2 M = C$, the coefficients and right-hand sides of the equation in property (iv) will satisfy uniform C^α-bounds. It follows from our previous estimate (2.4) and property (ii) that the functions $F^{(i)}$ satisfy uniform $C^{2,\alpha}$-bounds on the ball of radius 1/4 around the origin. A subsequence will then converge in C^2 to a limiting function G on $B_{1/4}$, with $|\partial^{\mathbf{l}}G(0)| = 1$ for some second-order partial derivative $\partial^{\mathbf{l}}$ by property (iii). But property (v) and the fact that $d_p^2 M = C \to \infty$ implies that G is identically zero on $B_{1/4}$, which is a contradiction. □

We will sometimes need the following strengthening of this estimate.

Proposition 2.9. *Under the same conditions as the previous theorem, we actually have a constant C such that*

$$\|f\|_{C^{k+2,\alpha}(\Omega')} \leqslant C(\|L(f)\|_{C^{k,\alpha}(\Omega)} + \|f\|_{L^1(\Omega)}).$$

To prove this, one just needs to show that under the conditions of Theorem 2.8, the C^0-norm of f is controlled by the L^1-norm of f together with the C^α-norm of Lf. This can be done by using a blow-up argument similar to what we have used above (a good exercise for the reader), although the more standard way is to use similar estimates in Sobolev spaces, together with the Sobolev embedding theorem. Note that in the special case when $Lf = 0$, this estimate generalizes the basic interior estimate corollary (Corollary 2.3) for harmonic functions.

An important point which does not follow from these arguments is that we do not need to know a priori that $f \in C^{k+2,\alpha}$. In other words if we just know that $f \in C^2$, so that the equation $L(f) = g$ makes sense, then if the coefficients of L and g are in $C^{k,\alpha}$, it follows that $f \in C^{k+2,\alpha}$ and inequality (2.3) holds. For this, one needs to work harder; see Gilbarg-Trudinger [**58**, Chapter 6].

On a smooth manifold M the Hölder spaces can be defined locally in coordinate charts. More precisely we cover M with coordinate charts U_i. Then any tensor T on M can be written in terms of its components on each U_i. The $C^{k,\alpha}$-norm of the tensor T can be defined as the supremum of the $C^{k,\alpha}$-norms of the components of T over each coordinate chart.

This works well if there are finitely many charts, which we can achieve if M is compact, for example. It is more natural, however, to work on Riemannian manifolds and define the Hölder norms relative to the metric. If (M, g) is a Riemannian manifold, then we can use parallel translation along geodesics with respect to the Levi-Civita connection to compare tensors at different points. For a tensor T we can define

$$|T|_{C^\alpha} = \sup_{x,y} \frac{|T(x) - T(y)|}{d(x,y)^\alpha},$$

where the supremum is taken over those pairs of points x, y which are connected by a unique minimal geodesic. The difference $T(x) - T(y)$ is computed by parallel transporting $T(y)$ to x along this minimal geodesic. We then define

$$\|f\|_{C^{k,\alpha}} = \sup_M (|f| + |\nabla f| + \cdots + |\nabla^k f|) + |\nabla^k f|_{C^\alpha}.$$

These Hölder norms are uniformly equivalent to the norms defined using charts, as long as we only have finitely many charts.

A linear second-order elliptic operator on a smooth manifold is an operator which in each local chart can be written as (2.1), where (a_{jk}) is symmetric and positive definite. The local Schauder estimates of Theorem 2.8 can easily be used to deduce global estimates on a compact manifold. In fact if we cover the manifold by coordinate charts U_i, then we will get estimates on slightly smaller open sets U_i', but we can assume that these still cover the manifold. We therefore obtain the following.

Theorem 2.10 (Schauder estimates). *Let (M, g) be a compact Riemannian manifold, and let L be a second-order elliptic operator on M. For any k and $\alpha \in (0, 1)$ there is a constant C such that*

$$\|f\|_{C^{k+2,\alpha}(M)} \leqslant C(\|L(f)\|_{C^{k,\alpha}(M)} + \|f\|_{L^1(M)}),$$

where C depends on (M, g), k, α, the $C^{k,\alpha}$-norms of the coefficients of L, and the constants of ellipticity λ, Λ in (2.2). As we mentioned above, it is enough to assume that $f \in C^2$, and it follows that actually $f \in C^{k+2,\alpha}$ whenever $L(f)$ and the coefficients of L are in $C^{k,\alpha}$.

This theorem has the important consequence that the solution spaces of linear elliptic equations on compact manifolds are finite dimensional. In particular the following argument can also be used to show that the space of global holomorphic sections $H^0(M, L)$ is finite dimensional if L is a holomorphic line bundle over a compact complex manifold M.

Corollary 2.11. *Let L be a second-order elliptic operator on a compact Riemannian manifold M. Then the kernel of L*

$$\ker L = \{f \in L^2(M) \mid f \textit{ is a weak solution of } Lf = 0\}$$

is a finite-dimensional space of smooth functions.

Proof. We know from Theorem 2.6, applied locally in coordinate charts, that any weak solution of $Lf = 0$ is actually smooth. To prove that $\ker L$ is finite dimensional we will prove that the closed unit ball in $\ker L$ with respect to the L^2 metric is compact. Indeed, let $f_k \in \ker L$ be a sequence of functions such that $\|f_k\|_{L^2(M)} \leqslant 1$. By Hölder's inequality we then have $\|f_k\|_{L^1(M)} \leqslant C_1$ for some constant C_1. Applying the Schauder estimates we obtain a constant C_2 such that

$$\|f_k\|_{C^{2,\alpha}(M)} \leqslant C_2.$$

It follows that a subsequence of the f_k converge in C^2 to a function f. Since the convergence is in C^2, we have $f \in \ker L$ and also $\|f\|_{L^2(M)} \leqslant 1$. This shows that any sequence in the unit ball of $\ker L$ has a convergent subsequence, so this ball is compact. Thus $\ker L$ must be finite dimensional. □

2.4. The Laplace operator on Kähler manifolds

The Laplace operator is the fundamental second-order differential operator on a Riemannian manifold. On Kähler manifolds we will use one-half of the usual Riemannian Laplacian, which can be written in terms of local holomorphic coordinates as

$$\Delta f = g^{k\bar{l}}\nabla_k\nabla_{\bar{l}}f = g^{k\bar{l}}\partial_k\partial_{\bar{l}}f.$$

Recall that $\nabla_k(\partial/\partial\bar{z}^l) = 0$, so the expression using partial derivatives holds even if we are not using normal coordinates, in contrast to the Riemannian case. Rewriting the operator in local real coordinates, we find that the Laplace operator is elliptic.

A useful way to think of the Laplacian is as the operator $\Delta = -\overline{\partial}^*\overline{\partial}$, where

$$\overline{\partial}^* : \Omega^{0,1}M \to C^\infty(M)$$

is the formal adjoint of $\overline{\partial}$. If our manifold is compact, then this means that for any $(0,1)$-form α and function f we have

$$\int_M \langle\alpha, \overline{\partial}f\rangle\, dV = \int_M \langle\overline{\partial}^*\alpha, f\rangle\, dV, \tag{2.7}$$

where $\langle\cdot,\cdot\rangle$ is the natural Hermitian form induced by the metric and dV is the Riemannian volume form, so $dV = \dfrac{\omega^n}{n!}$. So in local coordinates

$$\langle\alpha, \overline{\partial}f\rangle = g^{k\bar{l}}\alpha_{\bar{l}}\overline{\partial_{\bar{k}}f} = g^{k\bar{l}}\alpha_{\bar{l}}\partial_k\overline{f},$$

while $\langle\overline{\partial}^*\alpha, f\rangle$ is just the product $(\overline{\partial}^*\alpha)\overline{f}$. An integration by parts shows that the relation (2.7) implies that $\overline{\partial}^*\alpha = -g^{k\bar{l}}\nabla_k\alpha_{\bar{l}}$, and so $-\overline{\partial}^*\overline{\partial}$ agrees with our operator Δ. Note that by using covariant derivatives we do not have to worry about differentiating the metric when we integrate by parts, and at the same time remember that $\partial_k\overline{f} = \nabla_k\overline{f}$.

The same idea works for arbitrary (p,q) forms, giving rise to the Hodge Laplacian $\Delta = -\overline{\partial}^*\overline{\partial} - \overline{\partial}\overline{\partial}^*$ (the term $\overline{\partial}\overline{\partial}^*$ is zero on functions). We can also do the same with ∂ instead of $\overline{\partial}$ and on Kähler manifolds both give rise to the same Laplace operator. The following existence result for the Poisson equation illustrates a typical method for solving linear elliptic equations.

Theorem 2.12. *Suppose that (M,ω) is a compact Kähler manifold, and let $\rho : M \to \mathbf{R}$ be smooth such that*

$$\int_M \rho\, dV = 0. \tag{2.8}$$

Then there exists a smooth function $f : M \to \mathbf{R}$ such that $\Delta f = \rho$ on M. (Condition (2.8) is necessary since an integration by parts shows that Δf has zero integral for any f.)

Sketch of proof. One approach to the proof is to solve a variational problem. Namely we look for a function f minimizing the functional

$$E(f) = \int_M \left(\frac{1}{2}|\nabla f|^2 + \rho f\right) dV,$$

subject to the constraint $\int_M f\,dV = 0$. Here $|\nabla f|^2 = g^{k\bar{l}}\nabla_k f \nabla_{\bar{l}} f$. A suitable function space to work on is the space L^2_1 of functions which have one weak derivative in L^2. Alternatively L^2_1 is the completion of the space of smooth functions on M with respect to the norm

$$\|f\|_{L^2_1} = \int_M (|\nabla f|^2 + |f|^2)\,dV.$$

Using the Poincaré inequality one shows that there are constants ε, C such that

$$E(f) \geqslant \varepsilon \|f\|_{L^2_1} - C$$

for all f with zero mean. A minimizing sequence is therefore bounded in L^2_1, so a subsequence will be weakly convergent in L^2_1 to a function F. The lower semicontinuity of the L^2_1-norm implies that F will be a minimizer of E, and the weak convergence shows that $\int_M F\,dV = 0$. Now considering the variation of E at this minimizer F, we find that F is a weak solution of $\Delta F = \rho$ (condition (2.8) is used here). Finally Theorem 2.6 implies that F is actually smooth. □

With more work and some tools from functional analysis, one can obtain the following quite general theorem, which describes the mapping properties of linear elliptic operators between Hölder spaces on compact manifolds.

Theorem 2.13. *Let L be an elliptic second-order operator with smooth coefficients on a compact Riemannian manifold M. For $k \geqslant 0$ and $\alpha \in (0,1)$ suppose that $\rho \in C^{k,\alpha}(M)$ and that $\rho \perp \ker L^*$ with respect to the L^2-product. Then there exists a unique $f \in C^{k+2,\alpha}$ with $f \perp \ker L$ such that $Lf = \rho$. In other words, L is an isomorphism*

$$L : (\ker L)^\perp \cap C^{k+2,\alpha} \to (\ker L^*)^\perp \cap C^{k,\alpha}.$$

Chapter 3

Kähler-Einstein Metrics

Recall that a Riemannian metric is Einstein if its Ricci tensor is proportional to the metric. In this section, we are interested in Kähler metrics which are also Einstein. In other words we would like to find Kähler metrics ω which satisfy the equation

$$\mathrm{Ric}(\omega) = \lambda\omega,$$

for some $\lambda \in \mathbf{R}$. By rescaling the metric, we can assume that we are in one of three cases, depending on the sign of λ:

$$\mathrm{Ric}(\omega) = -\omega, \quad \mathrm{Ric}(\omega) = 0, \text{ or } \quad \mathrm{Ric}(\omega) = \omega.$$

As we have seen, the Ricci form of a Kähler metric defines a characteristic class of the manifold, namely

$$c_1(M) = \frac{1}{2\pi}[\mathrm{Ric}(\omega)],$$

which is independent of the Kähler metric ω on M. It follows that in order to find a Kähler-Einstein metric on M, the class $c_1(M)$ must either be a negative, zero, or positive cohomology class. In addition if $c_1(M)$ is positive or negative, then we can only hope to find an Einstein metric in a Kähler class proportional to $c_1(M)$.

The first main goal in this chapter is to study the case of a compact Kähler manifold M with $c_1(M) < 0$. In this case there exists a Kähler-Einstein metric on M, according to the following theorem of Aubin [**8**] and Yau [**122**].

Theorem 3.1. *Let M be a compact Kähler manifold with $c_1(M) < 0$. Then there is a unique Kähler metric $\omega \in -2\pi c_1(M)$ such that* $\mathrm{Ric}(\omega) = -\omega$.

There are lots of manifolds with $c_1(M) < 0$ (see Exercise 1.31), so using this theorem, it is possible to construct many examples of Einstein manifolds.

Next we turn to the case when $c_1(M) = 0$, in which case the Yau-Calabi theorem (Theorem 1.24) implies that every Kähler class contains a Kähler-Einstein metric, which is necessarily Ricci flat. Finally we briefly discuss the case $c_1(M) > 0$, which has only recently been solved. The algebro-geometric obstructions that appear in this and the more general case of extremal metrics will be our subject of study in the remainder of the book.

The basic reference for this chapter is Yau [**122**], but there are many places where this material is explained, for instance Siu [**97**], Tian [**113**], or Błocki [**17**].

3.1. The strategy

Our goal is to prove Theorem 3.1. First we rewrite the equation in terms of Kähler potentials. Let ω_0 be any Kähler metric in the class $-2\pi c_1(M)$. By the $\partial\bar{\partial}$-lemma there is a smooth function F on M such that

$$\mathrm{Ric}(\omega_0) = -\omega_0 + \sqrt{-1}\partial\bar{\partial}F. \tag{3.1}$$

If $\omega = \omega_0 + \sqrt{-1}\partial\bar{\partial}\varphi$ is another Kähler metric in the same class, then

$$\mathrm{Ric}(\omega) = \mathrm{Ric}(\omega_0) - \sqrt{-1}\partial\bar{\partial}\log\frac{\omega^n}{\omega_0^n},$$

so in order to make sure that $\mathrm{Ric}(\omega) = -\omega$, we need

$$-\sqrt{-1}\partial\bar{\partial}\varphi = \sqrt{-1}\partial\bar{\partial}F - \sqrt{-1}\partial\bar{\partial}\log\frac{\omega^n}{\omega_0^n}.$$

This will certainly be the case if we solve the equation

$$(\omega_0 + \sqrt{-1}\partial\bar{\partial}\varphi)^n = e^{F+\varphi}\omega_0^n. \tag{3.2}$$

At this point we can deal with the uniqueness statement in Theorem 3.1.

Lemma 3.2. *On a compact Kähler manifold M there exists at most one metric $\omega \in -2\pi c_1(M)$ such that* $\mathrm{Ric}(\omega) = -\omega$.

Proof. This is a simple application of the maximum principle. Suppose that $\mathrm{Ric}(\omega_0) = -\omega_0$, so in (3.1) above we can take $F = 0$. If $\omega = \omega_0 + \sqrt{-1}\partial\bar{\partial}\varphi$ also satisfies $\mathrm{Ric}(\omega) = -\omega$, then from (3.2) we get

$$(\omega_0 + \sqrt{-1}\partial\bar{\partial}\varphi)^n = e^{\varphi}\omega_0^n.$$

Suppose that φ achieves its maximum at $p \in M$. In local coordinates at p we have

$$\det(g_{j\bar{k}} + \partial_j\partial_{\bar{k}}\varphi) = e^{\varphi}\det(g_{j\bar{k}}),$$

but at p the matrix $\partial_j\partial_{\bar{k}}\varphi$ is negative semidefinite, so

$$\det(g_{j\bar{k}} + \partial_j\partial_{\bar{k}}\varphi)(p) \leqslant \det(g_{j\bar{k}})(p).$$

It follows that $\varphi(p) \leqslant 0$. Since we assumed that φ achieves its maximum at p, we have $\varphi(x) \leqslant 0$ for all x. Looking at the minimum point of φ we similarly find that $\varphi(x) \geqslant 0$ for all x, so we must have $\varphi = 0$. It follows that $\omega = \omega_0$. □

We will solve the equation using the continuity method. This involves introducing a family of equations depending on a parameter t, which for $t = 1$ gives the equation we want to solve, but for $t = 0$ it simplifies to a simpler equation. We use the family

$$(*)_t \qquad \begin{aligned} &(\omega_0 + \sqrt{-1}\partial\bar{\partial}\varphi)^n = e^{tF+\varphi}\omega_0^n, \\ &\omega_0 + \sqrt{-1}\partial\bar{\partial}\varphi \text{ is a Kähler form} \end{aligned}$$

for $t \in [0,1]$. The proof of Theorem 3.1 then comprises three steps:

(1) We can solve $(*)_0$. This is clear since $\varphi = 0$ is a solution of $(*)_0$.

(2) If $(*)_t$ has a solution for some $t < 1$, then for all sufficiently small $\varepsilon > 0$ we can also solve $(*)_{t+\varepsilon}$. This will be a consequence of the implicit function theorem.

(3) If for some $s \in (0,1]$ we can solve $(*)_t$ for all $t < s$, then we can also solve $(*)_s$. This is the heart of the matter, requiring estimates for the solutions in Hölder spaces, to ensure that we can take a limit along a subsequence as $t \to s$.

Given these three statements, it follows that we can solve $(*)_1$, proving Theorem 3.1. We now prove statement (2).

Lemma 3.3. *Suppose that $(*)_t$ has a smooth solution for some $t < 1$. Then for all sufficiently small $\varepsilon > 0$ we can also find a smooth solution of $(*)_{t+\varepsilon}$.*

Proof. Let us define the operator

$$\begin{aligned} F : C^{3,\alpha}(M) \times [0,1] &\to C^{1,\alpha}(M) \\ (\varphi, t) &\mapsto \log\frac{(\omega_0 + \sqrt{-1}\partial\bar{\partial}\varphi)^n}{\omega_0^n} - \varphi - tF. \end{aligned}$$

By our assumption we have a smooth function φ_t such that $F(\varphi_t, t) = 0$ and $\omega_t = \omega_0 + \sqrt{-1}\partial\bar{\partial}\varphi_t$ is a Kähler form. We use this Kähler metric ω_t to define the Hölder norms on M. In order to apply the implicit function theorem, we need to compute the derivative of F in the φ direction, at the point (φ_t, t):

$$DF_{(\varphi_t,t)}(\psi, 0) = \frac{n\sqrt{-1}\partial\overline{\partial}\psi \wedge \omega_t^{n-1}}{\omega_t^n} - \psi = \Delta_t\psi - \psi,$$

where Δ_t is the Laplacian with respect to ω_t. Let us write $L(\psi) = \Delta_t\psi - \psi$. This linear operator has trivial kernel: if $\Delta_t\psi - \psi = 0$, then necessarily $\psi = 0$ since

$$\int_M |\psi|^2\, dV_t = \int_M \psi\Delta_t\psi\, dV_t = -\int_M |\nabla\psi|_t^2\, dV_t \leqslant 0,$$

where we have put the t subscripts to indicate that everything is computed with respect to ω_t. The operator L is also selfadjoint, so L^* has trivial kernel. It follows from Theorem 2.13 that L is an isomorphism

$$L : C^{3,\alpha}(M) \to C^{1,\alpha}(M).$$

The implicit function theorem then implies that for s sufficiently close to t there exist functions $\varphi_s \in C^{3,\alpha}(M)$ such that $F(\varphi_s, s) = 0$. For s sufficiently close to t this φ_s will be close enough to φ_t in $C^{3,\alpha}$ to ensure that $\omega_0 + \sqrt{-1}\partial\overline{\partial}\varphi_s$ is a positive form.

What remains for us to show is that φ_s is actually smooth. We know that

$$\log\frac{(\omega_0 + \sqrt{-1}\partial\overline{\partial}\varphi_s)^n}{\omega_0^n} - \varphi_s - sF = 0.$$

In local coordinates, if ω_0 has components $g_{j\bar{k}}$, then we can write the equation as

$$\log\det\left(g_{j\bar{k}} + \partial_j\partial_{\bar{k}}\varphi_s\right) - \log\det(g_{j\bar{k}}) - \varphi_s - sF = 0.$$

Since we already have $\varphi_s \in C^{3,\alpha}$, we can differentiate the equation, with respect to z^l, say. We get

$$(g_s)^{j\bar{k}}\left(\partial_l g_{j\bar{k}} + \partial_l\partial_j\partial_{\bar{k}}\varphi_s\right) - \partial_l \log\det(g_{j\bar{k}}) - \partial_l\varphi_s - s\partial_l F = 0,$$

where $(g_s)^{j\bar{k}}$ is the inverse of the metric $(g_s)_{j\bar{k}} = g_{j\bar{k}} + \partial_j\partial_{\bar{k}}\varphi_s$ and we are using summation convention. Rewriting this equation,

$$(g_s)^{j\bar{k}}\partial_j\partial_{\bar{k}}(\partial_l\varphi_s) - \partial_l\varphi_s = s\partial_l F + \partial_l\log\det(g_{j\bar{k}}) - (g_s)^{j\bar{k}}\partial_l g_{j\bar{k}}.$$

We think of this as a linear elliptic equation $E(\partial_l\varphi_s) = h$ for the function $\partial_l\varphi_s$, where

$$h = s\partial_l F + \partial_l\log\det(g_{j\bar{k}}) - (g_s)^{j\bar{k}}\partial_l g_{j\bar{k}}.$$

Since $\varphi_s \in C^{3,\alpha}$, the coefficients of the operator E are in $C^{1,\alpha}$, and $h \in C^{1,\alpha}$. It follows that $\partial_l\varphi_s \in C^{3,\alpha}$. Similarly $\partial_{\bar{l}}\varphi_s \in C^{3,\alpha}$ so it follows that $\varphi_s \in C^{4,\alpha}$. Repeating the same argument, we get that $\varphi_s \in C^{5,\alpha}$, and inductively we find that φ_s is actually smooth. This technique of linearizing the equation and obtaining better and better regularity is called bootstrapping.

An alternative approach would be to use the implicit function theorem in $C^{k,\alpha}$ for larger and larger k, and the uniqueness of the solution will imply that the φ_s we obtain is actually smooth. □

The main difficulty is in step (3) of the strategy, namely that if we can solve $(*)_t$ for all $t < s$, then we can take a limit and thereby also solve $(*)_s$. For this we need the following a priori estimates.

Proposition 3.4. *There exists a constant $C > 0$ depending only on M, ω_0, and F such that if φ_t satisfies $(*)_t$ for some $t \in [0,1]$, then*

$$(g_{j\bar{k}} + \partial_j\partial_{\bar{k}}\varphi_t) > C^{-1}(g_{j\bar{k}}),$$

where the $g_{j\bar{k}}$ are the components of ω_0 in local coordinates and the inequality for matrices means that the difference is positive definite. In addition

$$\|\varphi_t\|_{C^{3,\alpha}(M)} \leqslant C,$$

where the Hölder norm is measured with respect to the metric ω_0.

We will prove this in the next two sections. For now we will show how it implies statement (3) in the strategy.

Lemma 3.5. *Assume Proposition 3.4. Suppose that $s \in (0,1]$ and that we can solve $(*)_t$ for all $t < s$. Then we can also solve $(*)_s$.*

Proof. Take a sequence of numbers $t_i < s$ such that $\lim t_i = s$. This gives rise to a sequence of functions φ_i which satisfy

$$(\omega_0 + \sqrt{-1}\partial\bar{\partial}\varphi_i)^n = e^{t_iF+\varphi_i}\omega_0^n. \tag{3.3}$$

Proposition 3.4 implies that the φ_i are uniformly bounded in $C^{3,\alpha}$, so by Theorem 2.7, after choosing a subsequence, we can assume that the φ_i converge to a function φ in $C^{3,\alpha'}$ for some $\alpha' < \alpha$. This convergence is strong enough that we can take a limit of (3.3), so we obtain

$$(\omega_0 + \sqrt{-1}\partial\bar{\partial}\varphi)^n = e^{sF+\varphi}\omega_0^n.$$

In addition Proposition 3.4 implies that the metrics $\omega_0 + \sqrt{-1}\partial\bar{\partial}\varphi_i$ are all bounded below by a fixed positive definite metric, so the limit $\omega_0 + \sqrt{-1}\partial\bar{\partial}\varphi$ is also positive definite.

Now the same argument as in the proof of Lemma 3.3 can be used to prove that φ is actually smooth. Alternatively Proposition 3.4 could be strengthened to give uniform bounds on the $C^{k,\alpha}$-norms of φ_t for all k, and then repeating the previous argument (combined with uniqueness) we would obtain a smooth solution. □

3.2. The C^0- and C^2-estimates

What remains is to prove Proposition 3.4. To simplify notation, we will write the equation as

$$(\omega + \sqrt{-1}\partial\bar{\partial}\varphi)^n = e^{F+\varphi}\omega^n, \tag{3.4}$$

and we write $g_{j\bar{k}}$ for the components of the metric ω in local coordinates. We will later apply the results with tF replacing F.

Lemma 3.6. *If φ satisfies equation* (3.4), *then* $\sup_M |\varphi| \leqslant \sup_M |F|$.

Proof. This is essentially the same argument as the uniquenes statement, Lemma 3.2. Suppose that φ achieves its maximum at $p \in M$. Then in local coordinates, the matrix $\partial_j\partial_{\bar{k}}\varphi$ is negative semidefinite at p, so

$$\det(g_{j\bar{k}} + \partial_j\partial_{\bar{k}}\varphi)(p) \leqslant \det(g_{j\bar{k}})(p).$$

Using equation (3.4) we get $F(p) + \varphi(p) \leqslant 0$, so $\varphi(p) \leqslant -F(p)$. Since φ is maximal at p, this means that

$$\sup_M \varphi \leqslant -F(p) \leqslant \sup_M |F|.$$

Similarly looking at the minimum point of φ shows that $\sup_M |\varphi| \leqslant \sup_M |F|$. □

Next we would like to find an estimate for the second derivatives of φ. In fact we obtain something weaker, namely an estimate for $\Delta\varphi$, which will imply bounds for the mixed partial derivatives $\partial_j\partial_{\bar{k}}\varphi$. It will be useful to write

$$g'_{j\bar{k}} = g_{j\bar{k}} + \partial_j\partial_{\bar{k}}\varphi,$$

so then

$$g^{j\bar{k}}g'_{j\bar{k}} = n + \Delta\varphi.$$

One more useful notation is to write $\mathrm{tr}_g g' = g^{j\bar{k}}g'_{j\bar{k}}$ and $\mathrm{tr}_{g'} g = g'^{j\bar{k}}g_{j\bar{k}}$. We will also write Δ' for the Laplacian with respect to the metric g'. The key calculation is the following.

Lemma 3.7. *There exists a constant B depending on M and g such that*

$$\Delta' \log \mathrm{tr}_g g' \geqslant -B\mathrm{tr}_{g'} g - \frac{g^{j\bar{k}}R'_{j\bar{k}}}{\mathrm{tr}_g g'},$$

where $R'_{j\bar{k}}$ is the Ricci curvature of g'.

Proof. We will compute in normal coordinates for the metric g around a point $p \in M$. In addition we can assume that g' is diagonal at p since any Hermitian matrix can be diagonalized by a unitary transformation. In particular at the point p we have

$$\mathrm{tr}_g g' = \sum_i g'_{i\bar{i}}, \quad \mathrm{tr}_{g'} g = \sum_j g'^{j\bar{j}} = \sum_j \frac{1}{g'_{j\bar{j}}}.$$

We can then compute that at p

$$\begin{aligned}\Delta' \mathrm{tr}_g g' &= g'^{p\bar{q}} \partial_p \partial_{\bar{q}} (g^{j\bar{k}} g'_{j\bar{k}}) \\ &= g'^{p\bar{q}} (\partial_p \partial_{\bar{q}} g^{j\bar{k}}) g'_{j\bar{k}} + g'^{p\bar{q}} g^{j\bar{k}} \partial_p \partial_{\bar{q}} g'_{j\bar{k}} \\ &= g'^{p\bar{q}} (\partial_p \partial_{\bar{q}} g^{j\bar{k}}) g'_{j\bar{k}} - g'^{p\bar{q}} g^{j\bar{k}} R'_{j\bar{k}p\bar{q}} + g'^{p\bar{q}} g^{j\bar{k}} g'^{a\bar{b}} (\partial_j g'_{p\bar{b}})(\partial_{\bar{k}} g'_{a\bar{q}}).\end{aligned}$$

Using that g' is diagonal, we have

$$\begin{aligned}g'^{p\bar{q}} (\partial_p \partial_{\bar{q}} g^{j\bar{k}}) g'_{j\bar{k}} &= \sum_{p,j} g'^{p\bar{p}} g'_{j\bar{j}} \partial_p \partial_{\bar{p}} g^{j\bar{j}} \\ &\geqslant -B \sum_{p,j} g'^{p\bar{p}} g'_{j\bar{j}} \\ &= -B (\mathrm{tr}_{g'} g)(\mathrm{tr}_g g'),\end{aligned}$$

where B is the largest of the numbers $-\partial_p \partial_{\bar{p}} g^{j\bar{j}}$ (more geometrically $-B$ is a lower bound for the bisectional curvature of g). We also have $g'^{p\bar{q}} R'_{j\bar{k}p\bar{q}} = R'_{j\bar{k}}$, so

$$\Delta' \mathrm{tr}_g g' \geqslant -B(\mathrm{tr}_{g'} g)(\mathrm{tr}_g g') - g^{j\bar{k}} R'_{j\bar{k}} + \sum_{p,j,a} g'^{p\bar{p}} g'^{a\bar{a}} |\partial_j g'_{p\bar{a}}|^2. \tag{3.5}$$

Incorporating the logarithm, we have

$$\begin{aligned}\Delta' \log \mathrm{tr}_g g' &= \frac{\Delta' \mathrm{tr}_g g'}{\mathrm{tr}_g g'} - \frac{g'^{pq} (\partial_p \mathrm{tr}_g g')(\partial_{\bar{q}} \mathrm{tr}_g g')}{(\mathrm{tr}_g g')^2} \\ &\geqslant -B \mathrm{tr}_{g'} g - \frac{g^{j\bar{k}} R'_{j\bar{k}}}{\mathrm{tr}_g g'} + \frac{1}{\mathrm{tr}_g g'} \sum_{p,j,a} g'^{p\bar{p}} g'^{a\bar{a}} |\partial_j g'_{p\bar{a}}|^2 \\ &\quad - \frac{1}{(\mathrm{tr}_g g')^2} \sum_{p,a,b} g'^{p\bar{p}} (\partial_p g'_{a\bar{a}})(\partial_{\bar{p}} g'_{b\bar{b}}).\end{aligned} \tag{3.6}$$

Now using the Cauchy-Schwarz inequality twice we have

$$\begin{aligned}
\sum_{p,a,b} g'^{p\bar{p}}(\partial_p g'_{a\bar{a}})(\partial_{\bar{p}} g'_{b\bar{b}}) &= \sum_{a,b}\sum_p \sqrt{g'^{p\bar{p}}}(\partial_p g'_{a\bar{a}})\sqrt{g'^{p\bar{p}}}(\partial_{\bar{p}} g'_{b\bar{b}}) \\
&\leqslant \sum_{a,b}\left(\sum_p g'^{p\bar{p}}|\partial_p g'_{a\bar{a}}|^2\right)^{1/2}\left(\sum_q g'^{q\bar{q}}|\partial_q g'_{b\bar{b}}|^2\right)^{1/2} \\
&= \left(\sum_a\left(\sum_p g'^{p\bar{p}}|\partial_p g'_{a\bar{a}}|^2\right)^{1/2}\right)^2 \\
&= \left(\sum_a \sqrt{g'_{a\bar{a}}}\left(\sum_p g'^{p\bar{p}} g'^{a\bar{a}}|\partial_p g'_{a\bar{a}}|^2\right)^{1/2}\right)^2 \\
&\leqslant \left(\sum_a g'_{a\bar{a}}\right)\left(\sum_{b,p} g'^{p\bar{p}} g'^{b\bar{b}}|\partial_p g'_{b\bar{b}}|^2\right).
\end{aligned}$$

It follows that

$$\begin{aligned}
\frac{1}{(\mathrm{tr}_g g')^2}\sum_{p,a,b} g'^{p\bar{p}}(\partial_p g'_{a\bar{a}})(\partial_{\bar{p}} g'_{b\bar{b}}) &\leqslant \frac{1}{\mathrm{tr}_g g'}\sum_{a,p} g'^{p\bar{p}} g'^{a\bar{a}}|\partial_p g'_{a\bar{a}}|^2 \\
&\leqslant \frac{1}{\mathrm{tr}_g g'}\sum_{a,j,p} g'^{p\bar{p}} g'^{a\bar{a}}|\partial_p g'_{j\bar{a}}|^2
\end{aligned}$$

since in the last sum we are simply adding in some non-negative terms. Finally, using the Kähler condition $\partial_p g'_{j\bar{a}} = \partial_j g'_{p\bar{a}}$, we obtain the required inequality from (3.6). □

Lemma 3.8. *There is a constant C depending on M, ω, $\sup_M |F|$, and a lower bound for ΔF such that a solution φ of (3.4) satisfies*

$$C^{-1}(g_{j\bar{k}}) < (g_{j\bar{k}} + \partial_j\partial_{\bar{k}}\varphi) < C(g_{j\bar{k}}).$$

Proof. Using the notation $g'_{j\bar{k}} = g_{j\bar{k}} + \partial_j\partial_{\bar{k}}\varphi$ as before, equation (3.4) implies

$$-R'_{j\bar{k}} = \partial_j\partial_{\bar{k}}F + \partial_j\partial_{\bar{k}}\varphi - R_{j\bar{k}} = \partial_j\partial_{\bar{k}}F + g'_{j\bar{k}} - g_{j\bar{k}} - R_{j\bar{k}}. \tag{3.7}$$

Using Lemma 3.7 we get

$$\Delta' \log \mathrm{tr}_g g' \geqslant -B\mathrm{tr}_{g'} g + \frac{\Delta F + \mathrm{tr}_g g' - n - R}{\mathrm{tr}_g g'},$$

where R is the scalar curvature of g. The Cauchy-Schwarz inequality implies that

$$(\mathrm{tr}_g g')(\mathrm{tr}_{g'} g) \geqslant n^2,$$

and since we are assuming a bound from below on ΔF, we have a constant C such that

$$\Delta' \log \operatorname{tr}_g g' \geqslant -B\operatorname{tr}_{g'} g - C\operatorname{tr}_{g'} g.$$

Now observe that

$$\Delta' \varphi = g'^{j\bar{k}}\partial_j \partial_{\bar{k}} \varphi = g'^{j\bar{k}}(g'_{j\bar{k}} - g_{j\bar{k}}) = n - \operatorname{tr}_{g'} g.$$

It follows that for $A = B + C + 1$ we have

$$\Delta'(\log \operatorname{tr}_g g' - A\varphi) \geqslant \operatorname{tr}_{g'} g - An.$$

Now suppose that $\log \operatorname{tr}_g g' - A\varphi$ achieves its maximum at $p \in M$. Then

$$0 \geqslant \Delta'(\log \operatorname{tr}_g g' - A\varphi)(p) \geqslant \operatorname{tr}_{g'} g(p) - An,$$

so

$$\operatorname{tr}_{g'} g(p) \leqslant An. \tag{3.8}$$

Choose normal coordinates for g at p such that g' is diagonal at p. Then (3.8) implies that at p we have

$$\frac{1}{g'_{i\bar{i}}} = g'^{i\bar{i}} \leqslant An \tag{3.9}$$

for each i. But from equation (3.4) we know that at p

$$\prod_{i=1}^{n} g'_{i\bar{i}} = e^{F(p)+\varphi(p)} \leqslant C_1 \tag{3.10}$$

for some constant C_1 since we are assuming a bound on $\sup |F|$, from which Lemma 3.6 implies a bound on $\sup |\varphi|$. Now (3.9) and (3.10) imply that $g'_{i\bar{i}} \leqslant C_2$ for each i, for some constant C_2. In particular

$$\operatorname{tr}_g g'(p) \leqslant nC_2.$$

Since $\log \operatorname{tr}_g g' - A\varphi$ achieves its maximum at p, we have

$$\log \operatorname{tr}_g g'(x) - A\varphi(x) \leqslant \log \operatorname{tr}_g g'(p) - A\varphi(p) \leqslant \log(nC_2) - A\varphi(p)$$

for any $x \in M$, so since from Lemma 3.6 we can bound $\sup |\varphi|$, we have

$$\sup_M \log \operatorname{tr}_g g' \leqslant C_3$$

for some constant C_3. Now if at a point x we choose normal coordinates for g in which g' is diagonal, then we have an upper bound on $g'_{i\bar{i}}(x)$ for each i. Inequality (3.10) holds at x too, so we also obtain a lower bound on each $g'_{i\bar{i}}(x)$. These upper and lower bounds on the metric g' are exactly what we wanted to prove. □

3.3. The C^3- and higher-order estimates

In this section we will derive estimates for the third derivatives of φ satisfying equation (3.4). We will follow the calculation in Phong-Sesum-Sturm [**90**] which is a more streamlined version of the original proofs in [**122**], [**7**], or rather their parabolic analog. It is also possible to use more general techniques to obtain a $C^{2,\alpha}$-estimate given the estimate on $\partial_j\partial_{\bar{k}}\varphi$ in the previous section, namely the complex version of Evans-Krylov's theorem (see [**17**] or [**97**] for this approach).

It will be convenient to change our notation slightly. We will write $\hat{g}_{j\bar{k}}$ for the fixed background metric and $g_{j\bar{k}} = \hat{g}_{j\bar{k}} + \partial_j\partial_{\bar{k}}\varphi$. We will use the equation for the Ricci curvature (3.7), which we will simply write in the form

$$R_{j\bar{k}} = -g_{j\bar{k}} + T_{j\bar{k}}, \tag{3.11}$$

where $R_{j\bar{k}}$ is the Ricci curvature of the (unknown) metric g and $T_{j\bar{k}}$ is a fixed tensor. We will use the estimate from Lemma 3.8, so we know that there is a constant Λ such that

$$\Lambda^{-1}(\hat{g}_{j\bar{k}}) < (g_{j\bar{k}}) < \Lambda(\hat{g}_{j\bar{k}}). \tag{3.12}$$

We would like to estimate the mixed third derivatives of φ. Since we have already bounded the metric, it is equivalent to estimate the Christoffel symbols $\Gamma^i_{jk} = g^{i\bar{l}}\partial_j g_{k\bar{l}}$. It is more natural to work with tensors, however, so we will focus on the difference of Christoffel symbols

$$S^i_{jk} = \Gamma^i_{jk} - \widehat{\Gamma}^i_{jk}, \tag{3.13}$$

where $\widehat{\Gamma}^i_{jk}$ are the Christoffel symbols of the Levi-Civita connection of $\hat{g}$. The key calculation now is the following.

Lemma 3.9. *Suppose that g satisfies equation* (3.11) *and the bound* (3.12). *There is a constant C depending on M, T, $\hat{g}$, and Λ such that*

$$\Delta|S|^2 \geqslant -C|S|^2 - C,$$

where $|S|$ is the norm of the tensor S measured with the metric g and Δ is the g-Laplacian.

Proof. To simplify the notation we will suppress the metric g. We will be computing with the Levi-Civita connection of g, so this will not cause any problems. For instance we will write

$$|S|^2 = g^{j\bar{k}}g^{a\bar{b}}g_{p\bar{q}}S^p_{ja}\overline{S^q_{kb}} = S^p_{ja}\overline{S^p_{ja}},$$

where we are still summing over repeated indices (alternatively we are working at a point in coordinates such that g is the identity). We have

$$\begin{aligned}\Delta|S|^2 &= \nabla_p\nabla_{\bar{p}}(S^i_{jk}\overline{S^i_{jk}}) \\ &= (\nabla_p\nabla_{\bar{p}}S^i_{jk})\overline{S^i_{jk}} + S^i_{jk}(\overline{\nabla_{\bar{p}}\nabla_p S^i_{jk}}) \\ &\quad + (\nabla_p S^i_{jk})(\overline{\nabla_p S^i_{jk}}) + (\nabla_{\bar{p}}S^i_{jk})(\overline{\nabla_{\bar{p}}S^i_{jk}}) \\ &\geqslant (\nabla_p\nabla_{\bar{p}}S^i_{jk})\overline{S^i_{jk}} + S^i_{jk}(\overline{\nabla_{\bar{p}}\nabla_p S^i_{jk}})\end{aligned} \tag{3.14}$$

since the two terms in the third line are squares. Commuting derivatives, we have

$$\begin{aligned}\nabla_{\bar{p}}\nabla_p S^i_{jk} &= \nabla_p\nabla_{\bar{p}}S^i_{jk} + R_j{}^m{}_{p\bar{p}}S^i_{mk} + R_k{}^m{}_{p\bar{p}}S^i_{jm} - R_m{}^i{}_{p\bar{p}}S^m_{jk} \\ &= \nabla_p\nabla_{\bar{p}}S^i_{jk} + R_j^{\ m}S^i_{mk} + R_k^{\ m}S^i_{jm} - R_m^{\ i}S^m_{jk},\end{aligned}$$

where $R_j^{\ m} = g^{m\bar{k}}R_{j\bar{k}}$ is the Ricci tensor of g with an index raised. By equation (3.11) and our assumptions, the Ricci tensor is bounded, so

$$|\nabla_{\bar{p}}\nabla_p S^i_{jk}| \leqslant |\nabla_p\nabla_{\bar{p}}S^i_{jk}| + C_1|S|, \tag{3.15}$$

for some constant C_1. We also have

$$\begin{aligned}\nabla_p\nabla_{\bar{p}}S^i_{jk} &= \nabla_p\partial_{\bar{p}}(\Gamma^i_{jk} - \widehat{\Gamma}^i_{jk}) \\ &= -\nabla_p(R_j{}^i{}_{k\bar{p}} - \widehat{R}_j{}^i{}_{k\bar{p}}) \\ &= -\nabla_k R_j^{\ i} + \widehat{\nabla}_p\widehat{R}_j{}^i{}_{k\bar{p}} + (\nabla_p - \widehat{\nabla}_p)\widehat{R}_j{}^i{}_{k\bar{p}},\end{aligned}$$

where we used the Bianchi identity $\nabla_p R_j{}^i{}_{k\bar{p}} = \nabla_k R_j{}^i{}_{p\bar{p}} = \nabla_k R_j^{\ i}$ and $\widehat{\nabla}, \widehat{R}$ are the Levi-Civita connection and curvature tensor of $\hat{g}$. The difference in the connections $\nabla_p - \widehat{\nabla}_p$ is bounded by S from definition (3.13), and so we can bound the covariant derivative $\nabla_k R_j^{\ i}$ using equation (3.11). We get

$$|\nabla_p\nabla_{\bar{p}}S^i_{jk}| \leqslant C_2|S| + C_3,$$

for some constants C_2, C_3. Combining this with (3.15) and (3.14) we get

$$\Delta|S|^2 \geqslant -(C_4|S| + C_5)|S| = -C_4|S|^2 - C_5|S|,$$

from which the required result follows. □

We are now ready to prove the third-order estimate.

Lemma 3.10. *Suppose that g satisfies equation* (3.11) *and the bound* (3.12). *Then there is a constant C depending on $M, T, \hat{g}$, and Λ such that $|S| \leqslant C$.*

Proof. Inequality (3.5) from our earlier calculation now implies (in our changed notation) that

$$\Delta \mathrm{tr}_{\hat{g}} g \geqslant -C_1 + \varepsilon|S|^2,$$

for some constants $\varepsilon, C_1 > 0$, since we are assuming that g and $\hat{g}$ are uniformly equivalent. Using the previous lemma, we can then choose a large constant A such that

$$\Delta(|S|^2 + A\mathrm{tr}_{\hat{g}}g) \geqslant |S|^2 - C_2,$$

for some C_2. Suppose now that $|S|^2 + A\mathrm{tr}_{\hat{g}}g$ achieves its maximum at $p \in M$. Then

$$0 \geqslant |S|^2(p) - C_2,$$

so $|S|^2(p) \leqslant C_2$. Then at every other point $x \in M$ we have

$$|S|^2(x) \leqslant |S|^2(x) + A\mathrm{tr}_{\hat{g}}g(x) \leqslant |S|^2(p) + A\mathrm{tr}_{\hat{g}}g(p) \leqslant C_2 + C_3,$$

for some C_3, which is what we wanted to prove. □

We can finally prove Proposition 3.4, which completes the proof of Aubin and Yau's theorem (Theorem 3.1). We recall the statement.

Proposition 3.11. *There exists a constant $C > 0$ depending only on M, ω_0, and F (in the application to Theorem 3.1, F is computed from ω_0) such that if φ_t satisfies the equation*

$$(\omega_0 + \sqrt{-1}\partial\overline{\partial}\varphi_t)^n = e^{tF+\varphi_t}\omega_0^n$$

for some $t \in [0,1]$, then

$$(g_{j\bar{k}} + \partial_j\partial_{\bar{k}}\varphi_t) > C^{-1}(g_{j\bar{k}})$$

and

$$\|\varphi_t\|_{C^{3,\alpha}(M)} \leqslant C,$$

where the Hölder norm is measured with respect to the metric ω_0.

Proof. Lemmas 3.6 and 3.8 together show that $g_{j\bar{k}} + \partial_j\partial_{\bar{k}}\varphi$ is uniformly equivalent to $g_{j\bar{k}}$. Then Lemma 3.10 shows that we have an a priori bound on the mixed third derivatives $\partial_j\partial_{\bar{k}}\partial_l\varphi$ and $\partial_{\bar{j}}\partial_{\bar{k}}\partial_l\varphi$. In particular this gives C^α-bounds on $\partial_j\partial_{\bar{k}}\varphi$. Now we can use the same argument of differentiating the equation and using the Schauder estimates as in Lemma 3.3 to get an a priori bound on $\|\varphi\|_{C^{3,\alpha}}$. □

Exercise 3.12. Under the same assumptions as Lemma 3.9 show that there is a constant C such that

$$\Delta|\mathrm{Rm}|^2 \geqslant -C|\mathrm{Rm}|^3 - C|\mathrm{Rm}| + |\nabla\mathrm{Rm}|^2 + |\overline{\nabla}\mathrm{Rm}|^2,$$

where Rm is the curvature tensor of g, so

$$|\mathrm{Rm}|^2 = g^{i\bar{p}}g^{k\bar{r}}g^{s\bar{l}}g_{j\bar{q}}R_{i\ k\bar{l}}^{\ j}\overline{R_{p\ r\bar{s}}^{\ q}}$$

and $\nabla\mathrm{Rm} = \nabla_p R_{i\ k\bar{l}}^{\ j}$ and $\overline{\nabla}\mathrm{Rm} = \nabla_{\bar{p}} R_{i\ k\bar{l}}^{\ j}$. Using this, show that there is a constant C such that

$$\Delta|\mathrm{Rm}| \geqslant -C|\mathrm{Rm}|^2 - C,$$

and finally using an argument similar to Lemma 3.10 show that under the same assumptions $|\mathrm{Rm}| \leqslant C$ for some C.

Exercise 3.13. Generalize the previous exercise to higher-order derivatives of the curvature, $|\nabla^k \mathrm{Rm}|$. In this way one can obtain a priori bounds on higher derivatives of a solution φ of equation (3.2) without appealing to the Schauder estimates.

3.4. The case $c_1(M) = 0$

When the manifold M has vanishing first Chern class, then a Kähler-Einstein metric on M is necessarily Ricci flat. Given any metric ω on M, the Ricci form of ω is exact, so by the $\partial\bar{\partial}$-lemma there is a function F such that

$$\mathrm{Ric}(\omega) = \sqrt{-1}\partial\bar{\partial}F.$$

Arguing as in the beginning of Section 3.1, we see that for $\omega' = \omega + \sqrt{-1}\partial\bar{\partial}\varphi$ to be Ricci flat, we need to solve the equation

$$(\omega + \sqrt{-1}\partial\bar{\partial}\varphi)^n = e^F\omega^n.$$

A slight difference from earlier is that for this to be possible, we first need to normalize F by adding a constant. In fact by integrating both sides of the equation, we have

$$\int_M e^F\omega^n = \int_M (\omega + \sqrt{-1}\partial\bar{\partial}\varphi)^n = \int_M \omega^n,$$

where we used that

$$\begin{aligned}(\omega + \sqrt{-1}\partial\bar{\partial}\varphi)^n - \omega^n &= \sqrt{-1}\partial\bar{\partial}\varphi \wedge (\omega'^{n-1} + \omega'^{n-2}\wedge\omega + \cdots + \omega^{n-1}) \\ &= d(\sqrt{-1}\bar{\partial}\varphi \wedge (\omega'^{n-1} + \cdots + \omega^{n-1}))\end{aligned}$$

is exact, so the volume of M with respect to the two different metrics is equal. The following theorem completely answers the $c_1(M) = 0$ case.

Theorem 3.14 (Yau). *Let (M, ω) be a compact Kähler manifold, and let $F : M \to \mathbf{R}$ be a smooth function such that*

$$\int_M e^F\omega^n = \int_M \omega^n.$$

Then there is a smooth function $\varphi : M \to \mathbf{R}$, unique up to the addition of a constant, such that $\omega + \sqrt{-1}\partial\bar{\partial}\varphi$ is a positive form and

$$(\omega + \sqrt{-1}\partial\bar{\partial}\varphi)^n = e^F\omega^n.$$

The equation looks very similar to equation (3.2) that we had to solve when proving Theorem 3.1, but there is one crucial difference. It is now not possible to prove an a priori estimate for $\sup_M |\varphi|$ using the maximum principle as we did in Lemma 3.6 since the function φ does not appear on the right-hand side of the equation. Nevertheless one can estimate $\sup_M |\varphi|$ using more sophisticated arguments due to Yau [**122**]. We will follow the exposition of Błocki [**17**] of Yau's proof, with simplifications due to Kazdan, Bourguignon, and Aubin.

Proposition 3.15. *Suppose that $F, \varphi : M \to \mathbf{R}$ are smooth functions on a compact Kähler manifold (M, ω) such that $\omega - ddb\varphi$ is positive and*

$$(\omega - \sqrt{-1}\partial\overline{\partial}\varphi)^n = e^F \omega^n.$$

Then there is a constant C depending on (M, ω) and $\sup_M F$ such that

$$\sup_M \varphi - \inf_M \varphi < C.$$

Proof. The proof is based on a technique called Moser iteration, originally used in the context of linear equations; see [**58**, Theorem 8.15]. The method is to estimate L^p-norms

$$\|\varphi\|_p = \left(\int_M |\varphi|^p \,\omega^n \right)^{1/p}$$

for higher and higher p iteratively and then take a limit as $p \to \infty$. Using $\omega - \sqrt{-1}\partial\overline{\partial}\varphi$ instead of $\omega + \sqrt{-1}\partial\overline{\partial}\varphi$ removes several negative signs in the arguments below. Modifying φ by a constant and rescaling ω, we can assume that $\inf_M \varphi = 1$ and $\int_M \omega^n = 1$. This ensures that $\|\varphi\|_p \leqslant \|\varphi\|_q$ for $p \leqslant q$. We will write C for a constant that may change from line to line but is only dependent on (M, ω) and $\sup_M F$.

The fact that $\omega - \sqrt{-1}\partial\overline{\partial}\varphi$ is positive implies, after we take the trace with respect to ω, that

$$n - \Delta\varphi > 0,$$

where Δ is the Laplacian of (M, ω). Suppose that $\varphi(p) = 1$, and let $G(x, y)$ be the Green's function of the Laplacian (see [**10**, Section 2.3]), so

$$\varphi(p) = \int_M \varphi\,\omega^n - \int_M G(x,p)\Delta\varphi(x)\,\omega^n(x).$$

We can assume that $G \geqslant 0$ and also that $G(x, p)$ is integrable with respect to x. It follows that

$$1 = \varphi(p) \geqslant \int_M \varphi\,\omega^n - n\int_M G(x,p)\,\omega^n(x) \geqslant \int_M \varphi\,\omega^n - C$$

for some constant C, so we get $\|\varphi\|_1 < C$.

Let us write $\omega_\varphi = \omega - \sqrt{-1}\partial\overline{\partial}\varphi$. We have

$$\begin{aligned}\int_M \varphi(\omega_\varphi^n - \omega^n) &= \int_M \varphi(\omega_\varphi - \omega) \wedge (\omega_\varphi^{n-1} + \cdots + \omega^n)\\ &= \int_M -\varphi\sqrt{-1}\partial\overline{\partial}\varphi \wedge (\omega_\varphi^{n-1} + \cdots + \omega^{n-1})\\ &= \int_M \sqrt{-1}\partial\varphi \wedge \overline{\partial}\varphi \wedge (\omega_\varphi^{n-1} + \cdots + \omega^{n-1}).\end{aligned}$$

The forms $\sqrt{-1}\partial\varphi\wedge\overline{\partial}\varphi\wedge\omega_\varphi^k\wedge\omega^{n-1-k}$ are all non-negative. This can be seen by calculating in coordinates at a point, where both ω and ω_φ are diagonal (see also Lemma 4.7). It follows that

$$\int_M \varphi(\omega_\varphi^n - \omega^n) \geqslant \int_M \sqrt{-1}\partial\varphi \wedge \overline{\partial}\varphi \wedge \omega^{n-1} = \frac{1}{n}\int_M |\partial\varphi|^2\omega^n,$$

where we used Lemma 4.7. Since $\omega_\varphi^n - \omega^n = (e^F - 1)\omega^n$, we have

$$\int_M |\partial\varphi|^2\omega^n < C,$$

for some constant C. The Poincaré inequality (see [**10**, Corollary 4.3]) on (M, ω) implies that

$$\int_M (\varphi - \|\varphi\|_1)^2\,\omega^n \leqslant C\int_M |\partial\varphi|^2\omega^n,$$

and so our previous bound on $\|\varphi\|_1$ now implies that $\|\varphi\|_2 < C$.

A similar calculation gives, for any $p \geqslant 2$, that

$$\begin{aligned}\int_M \varphi^{p-1}(\omega_\varphi^n - \omega^n) &= \int_M -\varphi^{p-1}\sqrt{-1}\partial\overline{\partial}\varphi \wedge (\omega_\varphi^{n-1} + \cdots + \omega^{n-1})\\ &= \int_M (p-1)\sqrt{-1}\varphi^{p-2}\partial\varphi \wedge \overline{\partial}\varphi \wedge (\omega_\varphi^{n-1} + \cdots + \omega^{n-1})\\ &= \frac{4(p-1)}{p^2}\int_M \sqrt{-1}\partial\varphi^{\frac{p}{2}} \wedge \overline{\partial}\varphi^{\frac{p}{2}} \wedge (\omega_\varphi^{n-1} + \cdots + \omega^{n-1})\\ &\geqslant \frac{4(p-1)}{np^2}\int_M |\partial\varphi^{\frac{p}{2}}|^2\,\omega^n,\end{aligned}$$

so we obtain

$$\|\partial\varphi^{\frac{p}{2}}\|_2^2 \leqslant Cp\|\varphi\|_{p-1}^{p-1},$$

for some constant C which is independent of p. The Sobolev inequality (see [**10**, Theorem 2.20]) for (M, ω) says that for any f we have

$$\|f\|_{\frac{2n}{n-1}}^2 \leqslant C_S\left(\|f\|_2^2 + \|\partial f\|_2^2\right),$$

for some constant C_S depending on (M,ω). Applying this to $f = \varphi^{\frac{p}{2}}$ we get

$$\begin{aligned}\|\varphi\|^p_{\frac{np}{n-1}} = \|\varphi^{\frac{p}{2}}\|^2_{\frac{2n}{n-1}} &\leqslant C_S\left(\|\varphi^{\frac{p}{2}}\|_2^2 + \|\partial\varphi^{\frac{p}{2}}\|_2^2\right)\\ &\leqslant C_S\left(\|\varphi\|_p^p + Cp\|\varphi\|_{p-1}^{p-1}\right)\\ &\leqslant Cp\|\varphi\|_p^p,\end{aligned}$$

and so

$$\|\varphi\|_{\frac{np}{n-1}} \leqslant (Cp)^{1/p}\|\varphi\|_p.$$

Writing $p_k = \left(\frac{n}{n-1}\right)^k p$, we get

$$\|\varphi\|_{p_k} \leqslant (Cp_{k-1})^{1/p_{k-1}}\|\varphi\|_{p_{k-1}} \leqslant \cdots \leqslant \|\varphi\|_p \prod_{i=0}^{k-1}(Cp_i)^{1/p_i} \leqslant \|\varphi\|_p \prod_{i=0}^{\infty}(Cp_i)^{1/p_i},$$

where the latter product is finite. Choosing $p = 2$ and letting $k \to \infty$, we get

$$\sup_M \varphi \leqslant C\|\varphi\|_2,$$

so our bound on the L^2-norm of φ implies the required bound on the supremum. □

Once we have an estimate for $\sup_M |\varphi|$, we can obtain higher-order estimates in exactly the same way as was done in Lemmas 3.8 and 3.10. The "openness" argument of Lemma 3.3 also goes through with minor changes, so the equation in Theorem 3.14 can be solved using the continuity method.

Exercise 3.16. Suppose that (M,ω) is a compact Kähler manifold and $\varphi : M \to \mathbf{R}$ is such that

$$(\omega + \sqrt{-1}\partial\bar{\partial}\varphi)^n = \omega^n.$$

Show that φ must be a constant, by using the identity

$$\int_M \varphi[(\omega + \sqrt{-1}\partial\bar{\partial}\varphi)^n - \omega^n] = 0$$

and integrating by parts. This proves the uniqueness statement in Theorem 3.14.

Exercise 3.17. Use Yau's theorem (Theorem 3.14) to prove the Calabi-Yau theorem (Theorem 1.24).

3.5. The case $c_1(M) > 0$

The remaining case is when $c_1(M) > 0$. Suppose that $\omega \in 2\pi c_1(M)$ is any Kähler metric. We are now seeking a metric $\omega' = \omega + \sqrt{-1}\partial\bar{\partial}\varphi$ such that $\mathrm{Ric}(\omega') = \omega'$. Arguing as at the beginning of Section 3.1, this requires solving the equation

$$(\omega + \sqrt{-1}\partial\bar{\partial}\varphi)^n = e^{F-\varphi}\omega^n.$$

In attempting to use the continuity method, the first problem is coming up with a family of equations for which we can show openness. Aubin [**9**] introduced the equations

$$\mathrm{Ric}(\omega_\varphi) = t\omega_\varphi + (1-t)\omega, \tag{3.16}$$

where $\omega_\varphi = \omega + \sqrt{-1}\partial\bar{\partial}\varphi$ and ω is a fixed Käher form in $2\pi c_1(M)$. If we write $\mathrm{Ric}(\omega) - \omega = \sqrt{-1}\partial\bar{\partial}F$, then this equation is equivalent to

$$(\omega + \sqrt{-1}\partial\bar{\partial}\varphi)^n = e^{F-t\varphi}\omega^n. \tag{3.17}$$

For $t = 0$ there is a solution to this equation by Yau's theorem (Theorem 1.24). When showing openness at $t = 0$, a slight technical difficulty is that the solution to the equation when $t = 0$ is not unique since we can add a constant to φ. This is reflected in the linearized operator not being invertible when $t = 0$. A simple way to overcome this issue is to fix a point $p \in M$ and solve the equations

$$(\omega + \sqrt{-1}\partial\bar{\partial}\varphi)^n = e^{F-t\varphi+\varphi(p)}\omega^n \tag{3.18}$$

instead. Then $\varphi - t^{-1}\varphi(p)$ will solve (3.17). As in the proof of Lemma 3.3 we can rewrite this equation as

$$\log\frac{\omega_\varphi^n}{\omega^n} + t\varphi - \varphi(p) - F = 0, \tag{3.19}$$

and at $t = 0$ the linearized operator at a solution φ is

$$\psi \mapsto \Delta_{\omega_\varphi}\psi - \psi(p),$$

which is an isomorphism $C^{3,\alpha}(M) \to C^{1,\alpha}(M)$. Indeed if $h \in C^{1,\alpha}(M)$ and $\bar{h}$ denotes the average of h with respect to ω_φ, then we can find a $\psi \in C^{3,\alpha}(M)$ solving

$$\Delta_{\omega_\varphi}\psi = h - \bar{h}.$$

We can then simply add a suitable constant to ψ in order to solve

$$\Delta_{\omega_\varphi}\psi - \psi(p) = h.$$

The implicit function theorem then implies that we can solve equation (3.19) for small $t > 0$. The openness at $t > 0$ is due to Aubin [**9**].

Lemma 3.18. *Suppose that φ is a solution of equation* (3.17) *for $t = s$, where $s \in (0,1)$. Then we can solve* (3.17) *for any t sufficiently close to s.*

Proof. To use the implicit function theorem, we simply need to show that the linearized operator is invertible. Rewriting (3.17) as

$$\log \frac{\omega_\varphi^n}{\omega^n} + t\varphi - F = 0,$$

the linearization of the operator at φ when $t = s$ is given by

$$L(\psi) = \Delta_{\omega_\varphi}\psi + s\psi.$$

In other words we need to show that the smallest non-zero eigenvalue of $-\Delta_{\omega_\varphi}$ is at least s, and for this the crucial input is that ω_φ satisfies

$$\mathrm{Ric}(\omega_\varphi) = s\omega_\varphi + (1-s)\omega,$$

i.e. the Ricci curvature of ω_φ is bounded below by s. The result essentially follows from the Bochner-Weitzenbock formula

$$\Delta_{\overline{\partial}} = \overline{\nabla}^*\overline{\nabla} + \mathrm{Ric}$$

for the $\overline{\partial}$-Laplacian acting on $(0,1)$-forms (see the proof of Lemma 7.7 for a generalization). More explicitly, suppose that $L(\psi) = 0$. Then we can compute (for simplicity we will supress the metric ω_φ)

$$\begin{aligned}
\int_M s\nabla_j\psi\nabla_{\bar{j}}\psi\,\omega_\varphi^n &= \int_M -\nabla_j\nabla_p\nabla_{\bar{p}}\psi\nabla_{\bar{j}}\psi\,\omega_\varphi^n \\
&= \int_M (-\nabla_{\bar{p}}\nabla_p\nabla_j\psi\nabla_{\bar{j}}\psi + R^{q\bar{j}}\nabla_q\psi\nabla_{\bar{j}}\psi)\,\omega_\varphi^n \\
&= \int_M (\nabla_p\nabla_j\psi\nabla_{\bar{p}}\nabla_{\bar{j}}\psi + s\nabla_j\psi\nabla_{\bar{j}}\psi + (1-s)\omega^{q\bar{j}}\nabla_q\psi\nabla_{\bar{j}}\psi)\,\omega_\varphi^n \\
&\geqslant \int_M (s\nabla_j\psi\nabla_{\bar{j}}\psi + (1-s)\omega^{q\bar{j}}\nabla_q\psi\nabla_{\bar{j}}\psi)\,\omega_\varphi^n,
\end{aligned}$$

where $R^{q\bar{j}}$ is the Ricci curvature of ω_φ and $\omega^{q\bar{j}}$ denotes the components of the metric ω, with indices raised using ω_φ. This inequality can only hold if ψ is a constant, but then $L(\psi) = 0$ implies that $\psi = 0$. Since L is selfadjoint, it follows that it is invertible. □

As before, what remains is to show that the set of t for which we can solve (3.17) is closed, and for this we need a priori estimates. Once again we cannot use the maximum principle to obtain an estimate for $\sup_M|\varphi|$ because the sign of φ is reversed. If we had such an estimate, then the same arguments as before could be used to solve the equation. It turns out, however, that not every manifold with $c_1(M) > 0$ admits a Kähler-Einstein metric, so in fact the equation cannot always be solved. The first obstructions due to Matsushima [**83**] and Futaki [**55**] were based on the automorphism group of M, and in the case of complex surfaces these turned out to be sufficient by the work of Tian [**111**]. Later a much more subtle obstruction called K-stability was found by Tian [**112**] motivated by a conjecture due to Yau [**123**]. In the remainder of this book we will study these obstructions, in particular K-stability. Much of the theory applies to a larger class of metrics introduced by Calabi [**21**], called extremal metrics, and it is these metrics that we will start to study in the next chapter.

Very recently, Chen-Donaldson-Sun [**31**] have shown that in fact K-stability of a manifold M with $c_1(M) > 0$ is sufficient for the existence of a Kähler-Einstein metric on M. The proof is significantly more involved than the other two cases, and so we will only make a few remarks about it. Letting T be the supremum of those t for which there is a solution, one needs to understand the behavior of the metrics ω_t as $t \to T$. It turns out that this is easier to do if instead of equation (3.16), one works with a variant, where the form α is concentrated along a subvariety $D \subset M$. This will be the case if we have a metric ω_t on M, which is only smooth on $M \setminus D$ and satisfies $\mathrm{Ric}(\omega_t) = t\omega_t$ there and which has conical singularities along D with cone angle $2\pi t$. The advantage of studying these metrics is that they are Kähler-Einstein away from D, and there are deep results on the limiting behavior of families of Einstein metrics. The work of Chen-Donaldson-Sun [**32**], [**33**], [**34**] shows that either $T = 1$ and the ω_t converge to a Kähler-Einstein metric on M, or one can contradict the K-stability assumption by studying the limiting behavior of the ω_t. Roughly speaking, this limit is always a Fano manifold with mild singularities, which admits a Kähler-Einstein metric, with conical singularities if $T < 1$. A detailed discussion of these results is beyond the scope of this book, but we will make some further remarks about them in Section 7.6.

3.6. Futher reading

We have seen that prescribing the Ricci curvature of a Kähler metric is equivalent to solving the complex Monge-Ampère equation

$$(\omega + \sqrt{-1}\partial\bar{\partial}\varphi)^n = F\omega^n, \tag{3.20}$$

where ω is a Kähler metric on an n-dimensional compact Kähler manifold M and $F : M \to \mathbf{R}$ is a given positive, smooth function satisfying the equation

$$\int_M F\omega^n = \int_M \omega^n.$$

Because of the fundamental nature of this equation, it has been studied extensively, and it has extensions to many different settings. In this section we will give a brief overview of some of the recent work that has been done. First of all, the equation can be studied on complex manifolds locally (see for instance Caffarelli-Kohn-Nirenberg-Spruck [**20**] and Bedford-Taylor [**13**]), but we will only mention results on compact manifolds. Note however that the work of Bedford-Taylor is crucial even in this case, in order to make sense of the "Monge-Ampère operator" on the left-hand side of equation (3.20) when φ is only assumed to be a bounded plurisubharmonic function.

A first extension is to consider equation (3.20) with a more general right-hand side, in particular relaxing the positivity of F to $F \geqslant 0$ and allowing less regularity. While some results in this direction have already been obtained in Yau's original paper [**122**], perhaps the most important result in this direction is the work of Kołodziej [**68**]. One particular result is that if $F \geqslant 0$ and $F \in L^p(M, \omega^n)$ for some $p > 1$, then equation (3.20) has a continuous solution. This has been further extended recently by Kołodziej [**69**] to obtain Hölder continuity of the solution.

A further generalization of the problem, studied for instance by Eyssidieux-Guedj-Zeriahi [**52**], concerns the case when ω is not a Kähler form but rather is the only semipositive closed (1,1)-form. The result established in [**52**] is that if ω is such a semipositive form on a compact Kähler manifold M with $\int_M \omega^n > 0$ and $F \in L^p(M, \omega^n)$, then there is a bounded solution to (3.20). A further extension of this to "big" cohomology classes was given by Boucksom-Eyssidieux-Guedj-Zeriahi [**19**]. Semipositive forms arise typically as pullbacks $\omega = f^*\eta$ of positive forms η under a holomorphic map $f : M \to N$, and in particular f could be a resolution of singularities of a singular manifold N. Applying their theory to this setting, in [**52**] the authors establish the existence of certain singular Kähler-Einstein metrics on any projective manifold of general type. These are manifolds M for which $\dim H^0(K_M^d)$ grows at a rate of $d^{\dim M}$, which is satisfied for instance when $c_1(M) < 0$. This result is thus a generalization of Aubin and Yau's theorem (Theorem 3.1). See also Song-Tian [**99**] for an alternative approach to obtaining these singular Kähler-Einstein metrics. For a recent survey of these results and more, see Phong-Song-Sturm [**91**].

Leaving the Kähler world, equation (3.20) can be studied on any compact complex manifold, with ω being a Hermitian metric. In this case, for smooth positive F, Tosatti-Weinkove [**117**] showed that a solution exists by proving the analog of the C^0-estimate, Proposition 3.15. The higher-order estimates had been established earlier by Cherrier [**37**]. In a different direction one can study an analog of equation (3.20) on symplectic manifolds, roughly speaking, by prescribing the volume form of a symplectic form in a fixed cohomology class, compatible with a given almost complex structure. This study was initiated by Donaldson [**47**] with potential applications to symplectic geometry. In this case the problem of solving the equation is still open in general, but substantial progress has been made by Tosatti and Weinkove (see [**118**] for a survey).

A final direction that we will discuss is the parabolic version of Equation (3.20). This is the equation

$$\frac{\partial\varphi}{\partial t} = \log\frac{(\omega+\sqrt{-1}\partial\overline{\partial}\varphi)^n}{\omega^n} + F, \tag{3.21}$$

where now φ is a function on $[0,T)\times M$ for some (possibly infinite) T, and we are given the initial condition $\varphi(0,\cdot)$. The first results were due to Cao [**23**], showing that a solution to the equation exists for all t, and converges as $t\to\infty$ (up to adding a time dependent constant) to the solution ψ of the equation

$$(\omega+\sqrt{-1}\partial\overline{\partial}\psi)^n = e^{-F+c}\omega^n,$$

for a suitable constant c, whose existence is guaranteed by Yau's Theorem 3.14. The reason why Equation (3.21) is particularly interesting is that on the level of the metrics $\omega+\sqrt{-1}\partial\overline{\partial}\varphi$ it is closely related to the Ricci flow

$$\frac{\partial\omega}{\partial t} = -\mathrm{Ric}(\omega).$$

This equation was introduced by Hamilton [**64**] and has lead to spectacular results in differential geometry, most famously Perelman's proof [**89**] of the Poincaré conjecture. In complex geometry the long time behavior of the Ricci flow, continued through certain singularities using surgery, has close connections with the minimal model program in algebraic geometry. See for instance Song-Tian [**98**] for the case of complex surfaces, and Song-Weinkove [**100**] for a survey of recent work in the area.

Chapter 4

Extremal Metrics

Suppose that M is a compact Kähler manifold with a Kähler class $\Omega \in H^2(M, \mathbf{R})$. A natural question is to ask for a particularly nice metric representing the class Ω. In the previous section we have seen that if $c_1(M) < 0$ and $\Omega = -c_1(M)$, then M admits a unique Kähler-Einstein metric, while if $c_1(M) = 0$, then any Kähler class on M admits a unique Ricci flat metric. Extremal metrics, introduced by Calabi [**21**], are a natural generalization of these to arbitrary Kähler classes on compact Kähler manifolds. When they exist, extremal metrics are good candidates for being the "best" metrics in a given Kähler class. In this section we will introduce extremal metrics and study some of their basic properties, while later on we will study obstructions to their existence.

4.1. The Calabi functional

As above, suppose that M is a compact Kähler manifold and $\Omega \in H^2(M, \mathbf{R})$ is a Kähler class.

Definition 4.1. An extremal metric on M in the class Ω is a critical point of the functional

$$\mathrm{Cal}(\omega) = \int_M S(\omega)^2\, \omega^n,$$

for $\omega \in \Omega$, where $S(\omega)$ is the scalar curvature. This functional is called the Calabi functional.

The first important result is understanding the Euler-Lagrange equation characterizing extremal metrics. For a function $f : M \to \mathbf{R}$ on a Kähler manifold, let us write $\mathrm{grad}^{1,0} f = g^{j\bar{k}}\partial_{\bar{k}} f$. This is a section of $T^{1,0}M$, and it is (up to a factor of 2), the (1,0)-part of the Riemannian gradient of f.

Theorem 4.2. *A metric ω on M is extremal if and only if* $\mathrm{grad}^{1,0}S(\omega)$ *is a holomorphic vector field.*

Proof. First let us study the variation of the Calabi functional under variations of a Kähler metric in a fixed Kähler class. So let $\omega_t = \omega + t\sqrt{-1}\partial\overline{\partial}\varphi$, and we will compute the derivative of $\mathrm{Cal}(\omega_t)$ at $t=0$. We have

$$\frac{d}{dt}\bigg|_{t=0} \omega_t^n = n\sqrt{-1}\partial\overline{\partial}\varphi \wedge \omega^{n-1} = \Delta\varphi\,\omega^n,$$

and so

$$\frac{d}{dt}\bigg|_{t=0} \mathrm{Ric}(\omega_t) = -\sqrt{-1}\partial\overline{\partial}\Delta\varphi.$$

Using that $S(\omega_t) = g_t^{j\bar{k}}R_{t,j\bar{k}}$,where $R_{t,j\bar{k}}$ is the Ricci curvature of ω_t, we have

$$\begin{aligned}\frac{d}{dt}\bigg|_{t=0} S(\omega_t) &= -g^{j\bar{q}}(\partial_p\partial_{\bar{q}}\varphi)g^{p\bar{k}}R_{j\bar{k}} - \Delta^2\varphi \\ &= -\Delta^2\varphi - R^{\bar{k}j}\partial_j\partial_{\bar{k}}\varphi.\end{aligned}$$

Writing $S = S(\omega)$ for simplicity, it follows that

$$\begin{aligned}\frac{d}{dt}\bigg|_{t=0} \mathrm{Cal}(\omega) &= \int_M [-2S(\Delta^2\varphi + R^{\bar{k}j}\partial_j\partial_{\bar{k}}\varphi) + S^2\Delta\varphi]\omega^n \\ &= \int_M \varphi\big[-2\Delta^2 S - 2\nabla_j\nabla_{\bar{k}}(R^{\bar{k}j}S) + \Delta(S^2)\big]\omega^n.\end{aligned}$$

Using the Bianchi identity $\nabla_{\bar{k}}R^{\bar{k}j} = g^{j\bar{k}}\nabla_{\bar{k}}S$, we have

$$\begin{aligned}\frac{d}{dt}\bigg|_{t=0} \mathrm{Cal}(\omega) &= \int_M \varphi\big[-2\Delta^2 S - 2\nabla_j(Sg^{j\bar{k}}\nabla_{\bar{k}}S + R^{\bar{k}j}\nabla_{\bar{k}}S) + \Delta(S^2)\big]\omega^n \\ &= \int_M \varphi\big[-2\Delta^2 S - 2\nabla_j(R^{\bar{k}j}\nabla_{\bar{k}}S)\big]\omega^n.\end{aligned}$$

In particular if ω is an extremal metric, then this variation must vanish for every φ, so

$$\Delta^2 S + \nabla_j(R^{\bar{k}j}\nabla_{\bar{k}}S) = 0.$$

Commuting derivatives, for any function ψ we have

$$\begin{aligned}\Delta^2\psi + \nabla_j(R^{\bar{k}j}\nabla_{\bar{k}}\psi) &= g^{j\bar{k}}g^{p\bar{q}}\nabla_j\nabla_{\bar{k}}\nabla_p\nabla_{\bar{q}}\psi + \nabla_j(R^{\bar{k}j}\nabla_{\bar{k}}\psi) \\ &= g^{j\bar{k}}g^{p\bar{q}}\nabla_j\nabla_p\nabla_{\bar{k}}\nabla_{\bar{q}}\psi - g^{j\bar{k}}g^{p\bar{q}}\nabla_j(R^{\bar{m}}_{\bar{q}p\bar{k}}\nabla_{\bar{m}}\psi) \\ &\quad + \nabla_j(R^{\bar{k}j}\nabla_{\bar{k}}\psi) \\ &= g^{j\bar{k}}g^{p\bar{q}}\nabla_p\nabla_j\nabla_{\bar{k}}\nabla_{\bar{q}}\psi.\end{aligned}$$

It follows that if we write

$$\begin{aligned}\mathcal{D} : C^\infty(M, \mathbf{C}) &\to C^\infty(\Omega^{0,1}M \otimes \Omega^{0,1}M) \\ \psi &\mapsto \nabla_{\bar{k}}\nabla_{\bar{q}}\psi,\end{aligned}$$

then

$$\Delta^2\psi + \nabla_j(R^{\bar{k}j}\nabla_{\bar{k}}\psi) = \mathcal{D}^*\mathcal{D}\psi,$$

where $\mathcal{D}^*$ is the formal adjoint of $\mathcal{D}$. In particular if $\mathcal{D}^*\mathcal{D}S = 0$, then

$$0 = \int_M S\mathcal{D}^*\mathcal{D}S\,\omega^n = \int_M |\mathcal{D}S|^2\,\omega^n,$$

so $\mathcal{D}S = 0$. Using the metric to identify $\Omega^{0,1}M \cong T^{1,0}M$, the operator $\mathcal{D}$ can also be thought of as

$$\mathcal{D}(\psi) = \nabla_{\bar{k}}(g^{j\bar{q}}\nabla_{\bar{q}}\psi) = \nabla_{\bar{k}}(\mathrm{grad}^{1,0}\psi) = \bar{\partial}(\mathrm{grad}^{1,0}\psi)$$

since on the holomorphic tangent bundle $T^{1,0}M$ the $(0,1)$-part of the covariant derivative coincides with the usual antiholomorphic partial derivatives. Therefore $\mathcal{D}S = 0$ is equivalent to saying that $\mathrm{grad}^{1,0}S$ is holomorphic. □

Definition 4.3. The fourth-order operator that appeared in the previous proof,

$$\begin{aligned}\mathcal{D}^*\mathcal{D}\psi &= \Delta^2\psi + \nabla_j(R^{\bar{k}j}\nabla_{\bar{k}}\psi)\\ &= \Delta^2\psi + R^{\bar{k}j}\nabla_j\nabla_{\bar{k}}\psi + g^{j\bar{k}}\nabla_j S\nabla_{\bar{k}}\psi,\end{aligned}$$

is called the Lichnerowicz operator. We saw in the proof that on a compact Kähler manifold $\mathcal{D}^*\mathcal{D}\psi = 0$ if and only if $\mathrm{grad}^{1,0}\psi$ is holomorphic. Note that in general this is a complex operator unless S is constant. One must remember this when using the selfadjointness of $\mathcal{D}^*\mathcal{D}$. For instance for complex-valued functions f, g we have

$$\int_M (\mathcal{D}^*\mathcal{D}f)\bar{g}\,\omega^n = \int_M f\overline{\mathcal{D}^*\mathcal{D}g}\,\omega^n.$$

From the previous proof we obtain a useful description of the variation of the scalar curvature under a variation of the metric.

Lemma 4.4. *Suppose that $\omega_t = \omega + t\sqrt{-1}\partial\bar{\partial}\varphi$. Then the scalar curvature S_t of ω_t satisfies*

$$\begin{aligned}\left.\frac{d}{dt}\right|_{t=0} S_t &= -\mathcal{D}^*\mathcal{D}\varphi + g^{j\bar{k}}\nabla_j S\nabla_{\bar{k}}\varphi\\ &= -\overline{\mathcal{D}^*\mathcal{D}\varphi} + g^{j\bar{k}}\nabla_j\varphi\nabla_{\bar{k}}S.\end{aligned}$$

Proof. The first formula follows from the previous proof. The second one follows by taking the conjugate and noting that S_t is real. □

Example 4.5. The most important examples of extremal metrics are constant scalar curvature Kähler metrics, which we will abbreviate as cscK. In fact most compact Kähler manifolds admit no non-zero holomorphic vector fields at all, so on such manifolds an extremal metric necessarily has constant scalar curvature.

In particular Kähler-Einstein metrics have constant scalar curvature, so they are examples of extremal metrics. Conversely suppose that ω is a cscK metric and we are in a Kähler class where a Kähler-Einstein metric could exist, i.e. $c_1(M) = \lambda[\omega]$ for some λ. Then ω is in fact Kähler-Einstein. Indeed, if the scalar curvature S is constant, then

$$\overline{\partial}^* R_{j\bar{k}} = -g^{p\bar{k}}\nabla_p R_{j\bar{k}} = -\nabla_j S = 0,$$

so the Ricci form is harmonic. But $2\pi\lambda\omega$ is also a harmonic form in the same class, so we have $R_{j\bar{k}} = 2\pi\lambda g_{j\bar{k}}$.

We will see in Section 4.4 that there are also examples of extremal metrics which do not have constant scalar curvature.

Exercise 4.6. Let ω be an extremal metric on a compact Kähler manifold M. Use the implicit function theorem to show that there exists an extremal metric in every Kähler class on M which is sufficiently close to $[\omega]$. This is a theorem of LeBrun-Simanca [**71**]. At first you should assume that M has no holomorphic vector fields, which simplifies the problem substantially. For the general case it might help to study Section 8.5.

In the next section we will further study the interplay between holomorphic vector fields and extremal metrics. In the remainder of this section we will show that in the definition of extremal metrics, instead of taking the L^2-norm of the scalar curvature, we could equivalently have taken the L^2-norms of the Ricci or Riemannian curvatures. For this we first need the following.

Lemma 4.7. *Let α and β be $(1,1)$-forms, given in local coordinates by $\alpha = \sqrt{-1}\alpha_{j\bar{k}}dz^j \wedge d\bar{z}^k$ and $\beta = \sqrt{-1}\beta_{j\bar{k}}dz^j \wedge d\bar{z}^k$ such that $\alpha_{j\bar{k}}$ and $\beta_{j\bar{k}}$ are Hermitian matrices. If ω is a Kähler metric with components $g_{j\bar{k}}$, then*

$$n\alpha \wedge \omega^{n-1} = (\mathrm{tr}_\omega \alpha)\omega^n,$$

$$n(n-1)\alpha \wedge \beta \wedge \omega^{n-2} = \big[(\mathrm{tr}_\omega\alpha)(\mathrm{tr}_\omega\beta) - \langle\alpha,\beta\rangle_\omega\big]\omega^n,$$

where $\mathrm{tr}_\omega\alpha = g^{j\bar{k}}\alpha_{j\bar{k}}$ and $\langle\alpha,\beta\rangle_\omega = g^{j\bar{k}}g^{p\bar{q}}\alpha_{j\bar{q}}\beta_{p\bar{k}}$.

Proof. We will prove the second equality since the first follows by taking $\beta = \omega$. We compute in local coordinates at a point where g is the identity and α is diagonal. Then

$$\omega = \sqrt{-1}\sum_i g_{i\bar{i}}dz^i \wedge d\bar{z}^i,$$

so

$$\omega^{n-2} = (\sqrt{-1})^{n-2}(n-2)!\sum_{i<j} dz^1 \wedge d\bar{z}^1 \wedge \cdots \wedge \widehat{dz^i \wedge d\bar{z}^i}$$

$$\wedge \cdots \wedge \widehat{dz^j \wedge d\bar{z}^j} \wedge \cdots \wedge dz^n \wedge d\bar{z}^n,$$

where the hats mean that those terms are omitted. Also

$$\alpha \wedge \beta = (\sqrt{-1})^2 \sum_{i \neq j} \alpha_{i\bar{i}} \beta_{j\bar{j}} dz^i \wedge d\bar{z}^i \wedge dz^j \wedge d\bar{z}^j$$
$$+ \text{(terms involving } \beta_{j\bar{k}} \text{ with } j \neq k)$$

since α is diagonal. It follows that

$$\begin{aligned} n(n-1)\alpha \wedge \beta \wedge \omega^{n-2} &= (\sqrt{-1})^n n! \sum_{i \neq j} \alpha_{i\bar{i}} \beta_{j\bar{j}} dz^1 \wedge d\bar{z}^1 \wedge \cdots \wedge dz^n \wedge d\bar{z}^n \\ &= \left(\sum_{i \neq j} \alpha_{i\bar{i}} \beta_{j\bar{j}} \right) \omega^n \\ &= \left(\sum_{i,j} \alpha_{i\bar{i}} \beta_{j\bar{j}} - \sum_i \alpha_{i\bar{i}} \beta_{i\bar{i}} \right) \omega^n \\ &= [(\mathrm{tr}_\omega \alpha)(\mathrm{tr}_\omega \beta) - \langle \alpha, \beta \rangle_\omega] \, \omega^n. \end{aligned}$$

□

We can now compare the different functionals obtained by taking the L^2-norms of the Ricci and Riemannian curvatures.

Corollary 4.8. *There are constants C_1, C_2 depending on M and the Kähler class Ω such that if $\omega \in \Omega$, then*

$$\begin{aligned} \int_M S\, \omega^n &= 2n\pi c_1(M) \cup [\omega]^{n-1}, \\ \int_M |\mathrm{Ric}|^2 \, \omega^n &= \int_M S^2 \, \omega^n + C_1, \\ \int_M |\mathrm{Rm}|^2 \, \omega^n &= \int_M |\mathrm{Ric}|^2 \, \omega^n + C_2, \end{aligned}$$

where S, Ric, and Rm are the scalar, Ricci, and Riemannian curvatures of ω.

Proof. Let us write $\rho = \sqrt{-1} R_{j\bar{k}} dz^j \wedge d\bar{z}^k$ for the Ricci form of ω and $g_{j\bar{k}}$ for the local components of the metric ω. Applying the previous lemma, we have

$$\int_M S\, \omega^n = n \int_M \rho \wedge \omega^{n-1} = 2n\pi c_1(M) \cup [\omega]^{n-1}$$

since $\mathrm{tr}_\omega \rho = S$ and ρ is a closed form representing the cohomology class $2\pi c_1(M)$.

For the second identity we again apply the previous lemma:

$$\int_M (S^2 - |\mathrm{Ric}|^2)\omega^n = n(n-1) \int_M \rho \wedge \rho \wedge \omega^{n-2} = 4n(n-1)\pi^2 c_1(M)^2 \cup [\omega]^{n-2}$$

since $\langle \rho, \rho \rangle_\omega = |\mathrm{Ric}|^2$.

For the third equation, let us introduce the endomorphism-valued 2-form Θ_p^q defined by

$$\Theta_p^q = \sqrt{-1}R_{p\,j\bar{k}}^{\ q}dz^j \wedge d\bar{z}^k.$$

Applying the previous lemma we have

$$\begin{aligned} n(n-1)\Theta_p^q \wedge \Theta_q^p \wedge \omega^{n-2} &= \left(R_p^q R_q^p - g^{j\bar{b}}g^{a\bar{k}}R_{p\,j\bar{k}}^{\ q}R_{q\,a\bar{b}}^{\ p}\right)\omega^n \\ &= \left(|\mathrm{Ric}|^2 - |\mathrm{Rm}|^2\right)\omega^n. \end{aligned}$$

The (2,2)-form $\Theta_p^q \wedge \Theta_q^p$ is a closed form whose cohomology class is independent of the metric (in fact it is the characteristic class $4\pi^2 c_1(M)^2 - 8\pi^2 c_2(M)$), and therefore

$$\int_M (|\mathrm{Ric}|^2 - |\mathrm{Rm}|^2)\,\omega^n = C_2. \qquad \square$$

For us the most important point from the previous result is that the average scalar curvature

$$\hat{S} = \frac{2n\pi c_1(M) \cup [\omega]^{n-1}}{[\omega]^n}$$

only depends on M and the Kähler class $[\omega]$. Since

$$\int_M S(\omega)^2\,\omega^n = \int_M (S(\omega) - \hat{S})^2\,\omega^n + \int_M \hat{S}^2\,\omega^n,$$

if a cscK metric exists in a Kähler class, then it minimizes the Calabi functional. It turns out that more generally extremal metrics minimize the Calabi functional in their respective Kähler classes, but this is much harder to prove. See Donaldson [**46**] and Exercise 7.24 for the case of projective manifolds and Chen [**30**] for Kähler manifolds.

Remark 4.9. An important consequence of the previous result is that if ω is an extremal metric, then we have an estimate for the L^2-norm of the curvature of ω. This can be exploited to understand how a family of extremal metrics could degenerate in certain cases. See for example Chen-LeBrun-Weber [**35**] for an existence result based on a careful analysis of the possible "blow-up" behaviors.

4.2. Holomorphic vector fields and the Futaki invariant

As before, M is a compact Kähler manifold with Kähler metric ω. A holomorphic vector field is a holomorphic section of $T^{1,0}M$. We will focus our attention on those vector fields which can be written as $v^j = g^{j\bar{k}}\partial_{\bar{k}}f$ for a function f. It is natural to allow complex-valued functions too. Let us define

$$\mathfrak{h} := \{\text{holomorphic sections } v \text{ of } T^{1,0}M \\ \text{such that } v^j = g^{j\bar{k}}\partial_{\bar{k}}f \text{ for some } f : M \to \mathbf{C}\}.$$

We have seen that $v^j = g^{j\bar{k}}\partial_{\bar{k}}f \in \mathfrak{h}$ if and only if $\mathcal{D}^*\mathcal{D}f = 0$ and v^j determines f up to the addition of a constant. We call f a *holomorphy potential* for v. We can identify $\mathfrak{h}$ with the functions in $\ker \mathcal{D}^*\mathcal{D}$ which have integral zero. The space $\mathfrak{h}$ is independent of the choice of metric in the Kähler class $[\omega]$ because of the following.

Lemma 4.10. *Let us write $g_{\varphi,j\bar{k}} = g_{j\bar{k}} + \partial_j\partial_{\bar{k}}\varphi$ for some φ. If $v \in \mathfrak{h}$ and $v^j = g^{j\bar{k}}\partial_{\bar{k}}f$, then*

$$v^j = g_\varphi^{j\bar{k}}\partial_{\bar{k}}(f + v(\varphi)),$$

where $v(\varphi) = v^i\partial_i\varphi$ is the derivative of φ along v.

Proof. We have

$$g_{\varphi,j\bar{p}}v^j = (g_{j\bar{p}} + \partial_j\partial_{\bar{p}}\varphi)g^{j\bar{k}}\partial_{\bar{k}}f = \partial_{\bar{p}}f + \partial_{\bar{p}}(v^j\partial_j\varphi),$$

where we used that $\nabla_{\bar{p}}v^j = \partial_{\bar{p}}v^j = 0$ since v is holomorphic. Multiplying this equation by the inverse of g_φ we get the required result. □

Exercise 4.11. Show that the space $\mathfrak{h}$ is closed under the Lie bracket.

Remark 4.12. It turns out that $\mathfrak{h}$ consists of precisely those holomorphic vector fields which have a zero somewhere (see LeBrun-Simanca [**72**]), so $\mathfrak{h}$ does not even depend on the choice of Kähler class. We will also see this in Exercise 4.15 which gives yet another characterization of $\mathfrak{h}$ amongst the holomorphic vector fields.

Exercise 4.13. Show that if $c_1(M) = 0$, then $\mathfrak{h} = \{0\}$.

Exercise 4.14. Give an example of a compact Kähler manifold M and a holomorphic section v of $T^{1,0}M$ such that $v \notin \mathfrak{h}$.

Exercise 4.15. Let v be a holomorphic vector field. Show that $v \in \mathfrak{h}$ if and only if $\alpha(v) = 0$ for all holomorphic $(1,0)$-forms α.

Exercise 4.16. Suppose that M is a Fano manifold, i.e. $c_1(M) > 0$. Show that then $\mathfrak{h}$ is the space of all holomorphic vector fields on M.

Remark 4.17. It is often useful to think of sections of $T^{1,0}M$ as real vector fields. This can be achieved by identifying $T^{1,0}M$ with the real tangent bundle TM, mapping a vector field of type (1,0) to its real part. In local coordinates $z^i = x^i + \sqrt{-1}y^i$. In view of equation (1.4), this means that

$$\frac{\partial}{\partial z^i} \mapsto \frac{1}{2}\frac{\partial}{\partial x^i}, \qquad \sqrt{-1}\frac{\partial}{\partial z^i} \mapsto \frac{1}{2}\frac{\partial}{\partial y^i}.$$

We can then calculate that if $f = u + \sqrt{-1}v$ is the decomposition of f into its real and imaginary parts, then

$$g^{j\bar{k}}\partial_{\bar{k}}f \mapsto \frac{1}{2}(\mathrm{grad}\, u + J\mathrm{grad}\, v),$$

where grad is the usual Riemannian gradient and J is the complex structure. We will see in Section 5.1 that $J\mathrm{grad}\, v$ is the Hamiltonian vector field corresponding to v with respect to the symplectic form ω. It follows that if $v \in \mathfrak{h}$ has a purely imaginary holomorphy potential, then the real part of v is a Killing field. Conversely, if the real part of v is a Killing field, then $v^j = g^{j\bar{k}}\partial_{\bar{k}}f$ for a purely imaginary function f.

In view of the previous remark, let us denote by $\mathfrak{k} \subset \mathfrak{h}$ the vector fields in $\mathfrak{h}$ which correspond to Killing vector fields under the identification $T^{1,0}M = TM$. The following is a basic result about the Lie algebra $\mathfrak{h}$ on a cscK manifold (see Lichnerowicz [**76**]).

Proposition 4.18. *Suppose that ω is a cscK metric on M. Then $\mathfrak{h} = \mathfrak{k} \oplus J\mathfrak{k}$.*

Proof. Let $\ker_0 \mathcal{D}^*\mathcal{D}$ denote the elements in the kernel with zero integral. Under the identification

$$\begin{aligned} \ker_0 \mathcal{D}^*\mathcal{D} &\xrightarrow{=} \mathfrak{h} \\ f &\mapsto g^{j\bar{k}}\partial_{\bar{k}}f, \end{aligned}$$

the subspace $\mathfrak{k}$ corresponds to the purely imaginary functions. On the other hand, when ω has constant scalar curvature, then

$$\mathcal{D}^*\mathcal{D} = \Delta^2 + R^{\bar{k}j}\nabla_j\nabla_{\bar{k}}$$

is a real operator, and so $u + \sqrt{-1}v \in \ker \mathcal{D}^*\mathcal{D}$ for real functions u, v if and only if $u, v \in \ker \mathcal{D}^*\mathcal{D}$. □

Remark 4.19. Since $\mathfrak{k}$ generates a compact group of automorphisms, this result implies that if M admits a cscK metric, then the Lie algebra $\mathfrak{h}$ is reductive. This can be used to give examples of manifolds which do not admit a cscK metric. For example if $M = \mathrm{Bl}_p\mathbf{CP}^2$ is the blow-up of the projective plane at one point, then using Exercise 4.16 and Exercise 8.1, the Lie algebra $\mathfrak{h}$ can be identified with the holomorphic vector fields on $\mathbf{CP}^2$ which vanish at p. This latter Lie algebra can be identified with the 3×3 matrices of the form

$$\begin{pmatrix} 0 & * & * \\ 0 & * & * \\ 0 & * & * \end{pmatrix},$$

which is not reductive. It follows that M does not admit cscK metrics in any Kähler class. In Exercise 4.32 we will see that M does admit extremal metrics in every Kähler class.

Remark 4.20. Studying the automorphism group can also be used to find manifolds which do not admit extremal metrics in any Kähler class. The proposition shows that if M admits a cscK metric and $\mathfrak{h}$ is non-trivial, then the group of holomorphic automorphisms of M must contain a compact subgroup. The same holds if M admits an extremal metric since then $J\mathrm{grad}\, S$ is a holomorphic Killing field. Amongst other examples, Levine [**74**] showed that if M is a suitable 4-point blow-up of $\mathbf{CP}^1 \times \mathbf{CP}^1$ (blowing up the points $(p,p),(p,q),(p,r),(q,p)$ with p,q,r any three points on $\mathbf{CP}^1$), then the identity component of the automorphism group of M is $\mathbf{C}$, and so it has no compact subgroup. It follows that M cannot admit an extremal metric in any Kähler class.

The main point in Proposition 4.18 was that when ω is a cscK metric, then $\mathcal{D}^*\mathcal{D}$ is a real operator. In general we have

$$(\mathcal{D}^*\mathcal{D} - \overline{\mathcal{D}^*\mathcal{D}})\varphi = g^{j\bar{k}}(\nabla_j S \nabla_{\bar{k}}\varphi - \nabla_j\varphi\nabla_{\bar{k}}S).$$

If ω is an extremal metric, then $v_{\mathbf{S}} = g^{j\bar{k}}\partial_{\bar{k}}S$ is holomorphic and if in addition $v_f = g^{j\bar{k}}\partial_{\bar{k}}f \in \mathfrak{h}$, then we can compute that

$$[v_{\mathbf{S}}, v_f] = g^{p\bar{q}}\nabla_{\bar{q}}g^{j\bar{k}}(\nabla_j\varphi\nabla_{\bar{k}}S - \nabla_j S\nabla_{\bar{k}}\varphi).$$

Denote by $\mathfrak{h}_{\mathbf{S}} \subset \mathfrak{h}$ the subalgebra commuting with $v_{\mathbf{S}}$, and note that elements in $\mathfrak{k}$ commute with $v_{\mathbf{S}}$ since they correspond to Killing fields. Then the same proof as in Proposition 4.18 can be used to show that $\mathfrak{h}_{\mathbf{S}} = \mathfrak{k} \oplus J\mathfrak{k}$. A further refinement of this result is given in Calabi [**22**].

The following theorem, due to Futaki [**55**], gives an obstruction to finding cscK metrics in a Kähler class. It will turn out to be a first glimpse into the obstruction given by K-stability.

Theorem 4.21. *Let (M,ω) be a compact Kähler manifold. Let us define the functional $F : \mathfrak{h} \to \mathbf{C}$, called the Futaki invariant, by*

$$F(v) = \int_M f(S - \hat{S})\,\omega^n, \tag{4.1}$$

where f is a holomorphy potential for v and $\hat{S}$ is the average of the scalar curvature S. This functional is independent of the choice of metric in the Kähler class $[\omega]$. In particular if $[\omega]$ admits a cscK metric, then $F(v) = 0$ for all $v \in \mathfrak{h}$.

Proof. Suppose that $\omega + \sqrt{-1}\partial\bar{\partial}\varphi$ is another Kähler metric in $[\omega]$, and write $\omega_t = \omega + t\sqrt{-1}\partial\bar{\partial}\varphi$. Let

$$F_t(v) = \int_M f_t(S_t - \hat{S})\omega_t^n,$$

where f_t is a holomorphy potential for v with respect to ω_t and S_t is the scalar curvature of ω_t. Note that by Corollary 4.8 the average $\hat{S}$ is independent of t. It is enough to show that the derivative of $F_t(v)$ at $t = 0$ vanishes. By Lemma 4.10, we can choose f_t so that

$$\left.\frac{d}{dt}\right|_{t=0} f_t = v^j \partial_j \varphi = g^{j\bar{k}} \partial_{\bar{k}} f \partial_j \varphi,$$

and from the proofs of Theorem 4.2 and Lemma 4.4 we have

$$\begin{aligned}\left.\frac{d}{dt}\right|_{t=0} \omega_t^n &= \Delta\varphi\, \omega^n,\\ \left.\frac{d}{dt}\right|_{t=0} S_t &= -\overline{\mathcal{D}^*\mathcal{D}\varphi} + g^{j\bar{k}}\partial_j\varphi\partial_{\bar{k}} S.\end{aligned}$$

It follows that

$$\begin{aligned}\left.\frac{d}{dt}\right|_{t=0} F_t(v) &= \int_M \Big[g^{j\bar{k}}\partial_{\bar{k}} f \partial_j \varphi (S - \hat{S}) - f(\overline{\mathcal{D}^*\mathcal{D}\varphi} - g^{j\bar{k}}\partial_j\varphi\partial_{\bar{k}} S) \\ &\qquad\qquad + f(S - \hat{S})\Delta\varphi\Big]\, \omega^n \\ &= \int_M -f\overline{\mathcal{D}^*\mathcal{D}\varphi}\, \omega^n \\ &= -\int_M \varphi \mathcal{D}^*\mathcal{D} f\, \omega^n,\end{aligned}$$

after writing $\Delta\varphi = g^{j\bar{k}}\partial_{\bar{k}}\partial_j\varphi$ and integrating by parts. Using that f is a holomorphy potential, we have $\mathcal{D}^*\mathcal{D} f = 0$, so the result follows. □

To compute the Futaki invariant using the defining formula directly is impractical if not impossible in all but the simplest cases. Instead, it is possible to use a localization formula to compute $F(v)$ for a holomorphic vector field by studying the zero set of v (see Tian [**113**]). A third approach, which will be fundamental in the later developments, is that if M is a projective manifold, then the Futaki invariant can be computed algebro-geometrically. We will explain this in Section 7.4.

A useful corollary to the previous theorem is the following.

Corollary 4.22. *Suppose that ω is an extremal metric on a compact Kähler manifold M. If the Futaki invariant vanishes (relative to the Kähler class $[\omega]$), then ω has constant scalar curvature.*

Proof. Since ω is an extremal metric, the vector field $v^j = g^{j\bar{k}}\partial_{\bar{k}} S$ is in $\mathfrak{h}$. It follows that

$$0 = F(v) = \int_M S(S - \hat{S})\omega^n = \int_M (S - \hat{S})^2\, \omega^n,$$

so we must have $S = \hat{S}$, i.e. S is constant. □

4.3. The Mabuchi functional and geodesics

In this section we will see that cscK metrics have an interesting variational characterization, discovered by Mabuchi [**80**], which is different from being critical points of the Calabi functional. Moreover this variational point of view gives insight into when we can expect a cscK metric to exist.

As before, let (M, ω) be a compact Kähler manifold. Let us write

$$\mathcal{K} = \{\varphi : M \to \mathbf{R} \mid \varphi \text{ is smooth and } \omega + \sqrt{-1}\partial\overline{\partial}\varphi > 0\},$$

for the space of Kähler potentials for Kähler metrics in the class $[\omega]$. For any $\varphi \in \mathcal{K}$ we will write

$$\omega_\varphi = \omega + \sqrt{-1}\partial\overline{\partial}\varphi$$

for the corresponding Kähler metric, and we will put a φ subscript on other geometric quantities to indicate that they refer to this metric. The tangent space $T_\varphi\mathcal{K}$ at φ can be identified with the smooth real-valued functions $C^\infty(M)$. We can therefore define a 1-form α on $\mathcal{K}$ by letting

$$\alpha_\varphi(\psi) = \int_M \psi(\hat{S} - S_\varphi)\,\omega_\varphi^n.$$

We can check that this 1-form is closed. This boils down to differentiating $\alpha_\varphi(\psi)$ with respect to φ and showing that the resulting 2-tensor is symmetric. More precisely we need to compute

$$\left.\frac{d}{dt}\right|_{t=0} \alpha_{\varphi+t\psi_2}(\psi_1)$$

and show that it is symmetric in ψ_1 and ψ_2. We have

$$\begin{aligned}
\left.\frac{d}{dt}\right|_{t=0} &\alpha_{\varphi+t\psi_2}(\psi_1) \\
&= \int_M \left[\psi_1(\mathcal{D}_\varphi^*\mathcal{D}_\varphi\psi_2 - g_\varphi^{j\bar{k}}\partial_j S_\varphi \partial_{\bar{k}}\psi_2) + \psi_1(\hat{S} - S_\varphi)\Delta_\varphi\psi_2\right]\omega_\varphi^n \\
&= \int_M \left[\psi_1\mathcal{D}_\varphi^*\mathcal{D}_\varphi\psi_2 - (\hat{S} - S_\varphi)g^{j\bar{k}}\partial_j\psi_1\partial_{\bar{k}}\psi_2\right]\omega_\varphi^n.
\end{aligned}$$

Switching ψ_1 and ψ_2 amounts to taking the conjugate of the whole expression (using selfadjointness of the complex operator $\mathcal{D}^*\mathcal{D}$). The left-hand side of the equation is real, so it follows that the expression is symmetric in ψ_1 and ψ_2.

Since α is a closed form and $\mathcal{K}$ is contractible, there exists a function $\mathcal{M} : \mathcal{K} \to \mathbf{R}$ such that $d\mathcal{M} = \alpha$ which we can normalize so that $\mathcal{M}(0) = 0$. We could get a more explicit formula by integrating α along straight lines, but the variation of $\mathcal{M}$ is more transparent. To summarize, we have the following.

Proposition 4.23. *There is a functional $\mathcal{M} : \mathcal{K} \to \mathbf{R}$ such that the variation of $\mathcal{M}$ along a path $\varphi_t = \varphi + t\psi$ is given by*

$$\frac{d}{dt}\bigg|_{t=0} \mathcal{M}(\varphi_t) = \int_M \psi(\hat{S} - S_\varphi)\omega_\varphi^n, \tag{4.2}$$

where S_φ is the scalar curvature of the metric $\omega_\varphi = \omega + \sqrt{-1}\partial\bar{\partial}\varphi$. This is called the Mabuchi functional or the K-energy.

Exercise 4.24. Suppose that we define the Mabuchi functional as follows. For any $\varphi \in \mathcal{K}$ let φ_t be a path in $\mathcal{K}$ such that $\varphi_0 = 0$ and $\varphi_1 = \varphi$, and define

$$\mathcal{M}(\varphi) = \int_0^1 \int_M \dot{\varphi}_t(\hat{S} - S_t)\omega_t^n \, dt,$$

where $\omega_t = \omega + \sqrt{-1}\partial\bar{\partial}\varphi_t$ and S_t is the scalar curvature of ω_t. Check directly that this is well-defined, i.e. the integral is independent of the path φ_t that we choose connecting 0 and φ.

Note that since the variation of $\mathcal{M}$ in the direction of the constant functions vanishes, $\mathcal{M}$ actually descends to a functional on the space of Kähler metrics in $[\omega]$. Moreover it is clear that critical points of $\mathcal{M}$ are given by constant scalar curvature metrics. A modification of the Mabuchi functional has been introduced by Guan [**60**] whose critical points are extremal metrics.

Next we will show that $\mathcal{M}$ is a convex function on $\mathcal{K}$ if we endow $\mathcal{K}$ with a natural Riemannian metric, introduced by Mabuchi [**81**] (see also Semmes [**94**] and Donaldson [**43**]). Given two elements $\psi_1, \psi_2 \in T_\varphi\mathcal{K}$ in the tangent space at $\varphi \in \mathcal{K}$, we can define the inner product

$$\langle \psi_1, \psi_2 \rangle_\varphi = \int_M \psi_1 \psi_2 \, \omega_\varphi^n.$$

This defines a Riemannian metric on the infinite-dimensional space $\mathcal{K}$. Let us first compute the equation satisfied by geodesics.

Proposition 4.25. *A path $\varphi_t \in \mathcal{K}$ is a (constant speed) geodesic if and only if*

$$\ddot{\varphi}_t - |\partial\dot{\varphi}_t|_t^2 = \ddot{\varphi}_t - g_t^{j\bar{k}}\partial_j\dot{\varphi}_t\partial_{\bar{k}}\dot{\varphi}_t = 0,$$

where the dots mean t-derivatives and g_t is the metric $\omega + \sqrt{-1}\partial\bar{\partial}\varphi_t$.

Proof. A constant speed geodesic is a critical point of the energy of a path. The energy of the path φ_t for $t \in [0,1]$, say, is

$$E(\varphi_t) = \int_0^1 \int_M \dot{\varphi}_t^2 \, \omega_t^n \, dt.$$

Under a variation $\varphi_t + \varepsilon\psi_t$, where ψ_t vanishes at $t = 0$ and $t = 1$, we have

$$\begin{aligned}\frac{d}{d\varepsilon}\bigg|_{\varepsilon=0} E(\varphi_t + \varepsilon\psi_t) &= \int_0^1 \int_M (2\dot{\varphi}_t\dot{\psi}_t + \dot{\varphi}_t^2\Delta_t\psi_t)\,\omega_t^n\,dt \\ &= \int_0^1 \int_M (2\dot{\varphi}_t\dot{\psi}_t + \Delta(\dot{\varphi}_t^2)\psi_t)\,\omega_t^n\,dt \\ &= \int_0^1 \int_M \big[-2\ddot{\varphi}_t\psi_t - 2\dot{\varphi}_t\psi_t\Delta_t\dot{\varphi}_t + \Delta_t(\dot{\varphi}_t^2)\psi_t\big]\,\omega_t^n\,dt \\ &= \int_0^1 \int_M -2\psi_t\big[\ddot{\varphi}_t - g_t^{j\bar{k}}\partial_j\dot{\varphi}_t\partial_{\bar{k}}\dot{\varphi}_t\big]\,\omega_t^n\,dt,\end{aligned}$$

where we integrated by parts on the manifold and also with respect to t (the $\Delta_t\dot{\varphi}_t$ term in the third line comes from differentiating ω_t^n with respect to t). The required expression for the geodesic equation follows. □

Example 4.26. A useful family of geodesics arises as follows. Suppose that $v \in \mathfrak{h}$ has holomorphy potential $u : M \to \mathbf{R}$ and $v_{\mathbf{R}}$ is the real part of v, thought of as a section of TM. Then $v_{\mathbf{R}} = \frac{1}{2}\mathrm{grad}u$, and $v_{\mathbf{R}}$ is a real holomorphic vector field; i.e. the one-parameter group of diffeomorphisms $f_t : M \to M$ generated by $v_{\mathbf{R}}$ preserves the complex structure of M. We can then define the path of metrics

$$\omega_t = f_t^*(\omega),$$

and we can check that

$$\omega_t = \omega + \sqrt{-1}\partial\bar{\partial}\varphi_t,$$

where

$$\dot{\varphi}_t = f_t^* u.$$

A good exercise is to check that φ_t defines a geodesic line in $\mathcal{K}$. The derivative of the Mabuchi functional along this line is given by

$$\begin{aligned}\frac{d}{dt}\mathcal{M}(\varphi_t) &= \int_M \dot{\varphi}_t(\hat{S} - S_t)\,\omega_t^n \\ &= \int_M f_t^* u\,(\hat{S} - f^*S(\omega))\,f^*(\omega^n) \\ &= \int_M u(\hat{S} - S(\omega))\,\omega^n \\ &= -F(v),\end{aligned}$$

where $F(v)$ is the Futaki invariant of v. In other words the Mabuchi functional is linear along this geodesic line, with derivative given by the Futaki invariant.

Proposition 4.27. *The Mabuchi functional $\mathcal{M} : \mathcal{K} \to \mathbf{R}$ is convex along geodesics.*

Proof. Suppose that φ_t defines a geodesic and let us compute the second derivative of $\mathcal{M}(\varphi_t)$. By definition

$$\frac{d}{dt}\mathcal{M}(\varphi_t) = \int_M \dot{\varphi}_t(\hat{S} - S_t)\,\omega_t^n,$$

so

$$\begin{aligned}\frac{d^2}{dt^2}\mathcal{M}(\varphi_t) &= \int_M \left[\ddot{\varphi}_t(\hat{S} - S_t) + \dot{\varphi}_t(\mathcal{D}_t^*\mathcal{D}_t\dot{\varphi}_t - g_t^{j\bar{k}}\partial_j S_t \partial_{\bar{k}}\dot{\varphi}_t)\right.\\ &\qquad\qquad \left. + \dot{\varphi}_t(\hat{S} - S_t)\Delta_t\dot{\varphi}_t\right]\omega_t^n \\ &= \int_M \left[|\mathcal{D}_t\dot{\varphi}_t|_t^2 + (\hat{S} - S_t)(\ddot{\varphi}_t - |\partial\dot{\varphi}_t|_t^2)\right]\omega_t^n \qquad (4.3)\\ &= \int_M |\mathcal{D}_t\dot{\varphi}_t|_t^2\,\omega_t^n \geqslant 0.\end{aligned}$$

Therefore $\mathcal{M}$ is convex along the path φ_t. □

From this result a very appealing picture arises. We have a convex functional $\mathcal{M} : \mathcal{K} \to \mathbf{R}$ whose critical points are the cscK metrics in the class $[\omega]$. We can therefore at least heuristically expect a cscK metric to exist if and only if as we approach the "boundary" of $\mathcal{K}$, the derivative of $\mathcal{M}$ becomes positive. Since we are on an infinite-dimensional space, it is hard to make this picture rigorous, but we will find that the notion of K-stability can be seen as an attempt to encode this behavior "at infinity" of the functional $\mathcal{M}$.

Unfortunately it is difficult to construct geodesics in $\mathcal{K}$, and in fact it is possible to construct pairs of potentials in $\mathcal{K}$ on any Kähler manifold which are not joined by a smooth geodesic (see Lempert-Vivas [**73**], Darvas [**38**]). Nevertheless it is possible to show the existence of non-smooth geodesics with enough regularity that geometric conclusions can be drawn (see Chen [**29**], Chen-Tian [**36**]). In particular Chen and Tian showed that extremal metrics, if they exist, are unique up to isometry in a Kähler class.

Exercise 4.28. Suppose that ω_1, ω_2 are two cscK metrics in the same Kähler class on M. Assuming that there is a geodesic path connecting ω_1 and ω_2, prove that there is a biholomorphism $f : M \to M$ such that $f^*\omega_2 = \omega_1$.

To conclude this section, we briefly mention that when $\omega \in c_1(M)$, then there is another natural functional on $\mathcal{K}$ whose critical points are Kähler-Einstein metrics, introduced by Ding [**40**]. To define it, for any $\varphi \in \mathcal{K}$ define the Ricci potential h_φ by the equation

$$\mathrm{Ric}(\omega_\varphi) - \omega_\varphi = \sqrt{-1}\partial\bar{\partial}h_\varphi,$$

together with the normalization

$$\int_M e^{h_\varphi}\omega_\varphi^n = \int_M \omega_\varphi^n.$$

The variation of the Ding functional $\mathcal{F} : \mathcal{K} \to \mathbf{R}$ along a path $\varphi_t = \varphi + t\psi$ is given by

$$\left.\frac{d}{dt}\right|_{t=0} \mathcal{F}(\varphi_t) = \int_M \psi(e^{h_\varphi} - 1)\omega_\varphi^n.$$

Exercise 4.29. Show that if $\int_M f e^{h_\varphi}\omega_\varphi^n = 0$, then

$$\int_M f^2 e^{h_\varphi}\,\omega_\varphi^n \leqslant \int_M |\nabla f|^2 e^{h_\varphi}\omega_\varphi^n.$$

Exercise 4.30. Show that a functional $\mathcal{F}$ exists with the variational formula above and that $\mathcal{F}$ is convex along smooth geodesics in $\mathcal{K}$.

The advantage of the Ding functional $\mathcal{F}$ over the Mabuchi functional is that it can be defined for metrics with less regularity. In particular the convexity of $\mathcal{F}$ can be established along geodesics in $\mathcal{K}$ with very low regularity, and this leads to results on the uniqueness of Kähler-Einstein metrics, even ones with certain singularities, as shown by Berndtsson [**15**]. Note that the uniqueness of smooth Kähler-Einstein metrics up to isometry has previously been established by Bando-Mabuchi [**11**] without the use of geodesics.

4.4. Extremal metrics on a ruled surface

In this section we will describe the construction of explicit extremal metrics on a ruled surface, due to Tønnesen-Friedman [**116**]. We will only do the calculation in a special case, but much more general results along these lines can be found in the work of Apostolov–Calderbank–Gauduchon–Tønnesen-Friedman [**2**].

Let Σ be a genus 2 curve, and let ω_Σ be a Kähler metric on Σ with constant scalar curvature -2. By the Gauss-Bonnet theorem the area of Σ is 2π with this metric. Let L be a degree -1 holomorphic line bundle on Σ (i.e. $c_1(L)[\Sigma] = -1$), and let h be a metric on L with curvature form $F(h) = -\omega_\Sigma$.

We will construct metrics on the projectivization $X = \mathbf{P}(L \oplus \mathcal{O})$ over Σ, where $\mathcal{O}$ is the trivial line bundle. Thus X is a $\mathbf{CP}^1$-bundle over Σ. We will follow the method of Hwang-Singer [**65**]. First we construct metrics on the complement of the zero section in the total space of L and then describe what is necessary to complete the metrics across the zero and infinity sections of X.

We will consider metrics of the form

$$\omega = p^*\omega_\Sigma + \sqrt{-1}\partial\overline{\partial} f(s), \tag{4.4}$$

where $p : L \to \Sigma$ is the projection map, $s = \log |z|_h^2$, and f is a strictly convex function which makes ω positive definite. Let us compute the metric ω in local coordinates. Choose a local holomorphic coordinate z on Σ and a fiber coordinate w for L, corresponding to a holomorphic trivialization around z. The fiberwise norm is then given by $|(z,w)|_h^2 = |w|^2 h(z)$ for some function h, and so our coordinate s is given by

$$s = \log |w|^2 + \log h(z).$$

Let us work at a point (z_0, w_0), in a trivialization such that $d \log h(z_0) = 0$. Then at this point

$$\begin{aligned}\sqrt{-1}\partial\bar{\partial} f(s) &= f'(s)\sqrt{-1}\partial\bar{\partial}\log h + f''(s)\sqrt{-1}\frac{dw \wedge d\bar{w}}{|w|^2} \\ &= f'(s) p^*\omega_\Sigma + f''(s)\sqrt{-1}\frac{dw \wedge d\bar{w}}{|w|^2},\end{aligned} \tag{4.5}$$

where we used that $-\sqrt{-1}\partial\bar{\partial}\log h$ is the curvature of L. It follows that

$$\omega = (1 + f'(s)) p^*\omega_\Sigma + f''(s)\sqrt{-1}\frac{dw \wedge d\bar{w}}{|w|^2}, \tag{4.6}$$

and so

$$\omega^2 = \frac{1}{|w|^2}(1 + f'(s)) f''(s) p^*\omega_\Sigma \wedge (\sqrt{-1} dw \wedge d\bar{w}).$$

We can check that if we now use a different trivialization for the line bundle in which $\tilde{w} = g(z)w$ for a holomorphic function z, then the same formula for ω^2 holds, so this formula holds at every point. It follows that the Ricci form of ω is

$$\begin{aligned}\rho &= -\sqrt{-1}\partial\bar{\partial}\log\left(\frac{1}{|w|^2}(1 + f'(s)) f''(s)\right) + p^*\rho_\Sigma \\ &= -\sqrt{-1}\partial\bar{\partial}\log\Big[(1 + f'(s)) f''(s)\Big] - 2p^*\omega_\Sigma,\end{aligned} \tag{4.7}$$

where $\rho_\Sigma = -2\omega_\Sigma$ is the Ricci form of Σ. We could at this point compute the scalar curvature of ω, but it is more convenient to change coordinates. From (4.6) we know that for ω to be positive, f must be strictly convex. We can therefore take the Legendre transform of f. The Legendre transform F is defined in terms of the variable $\tau = f'(s)$ by the formula

$$f(s) + F(\tau) = s\tau.$$

If $I \subset \mathbf{R}$ is the image of f', then F is a strictly convex function defined on I. The momentum profile of the metric is defined to be $\varphi : I \to \mathbf{R}$, where

$$\varphi(\tau) = \frac{1}{F''(\tau)}.$$

The following relations can be verified:

$$s = F'(\tau), \quad \frac{ds}{d\tau} = F''(\tau), \quad \varphi(\tau) = f''(s).$$

Using (4.6) and (4.7) we have

$$\begin{aligned} \omega &= (1+\tau)p^*\omega_\Sigma + \varphi(\tau)\frac{\sqrt{-1}dw \wedge d\bar{w}}{|w|^2}, \\ \rho &= -\sqrt{-1}\partial\bar{\partial}\log\big[(1+\tau)\varphi(\tau)\big] - 2p^*\omega_\Sigma. \end{aligned} \tag{4.8}$$

A calculation now shows that the scalar curvature is given by

$$S(\tau) = -\frac{2}{1+\tau} - \frac{1}{1+\tau}\big[(1+\tau)\varphi\big]'', \tag{4.9}$$

where the primes mean derivatives with respect to τ.

We still need to understand when we can complete the metric across the zero and infinity sections. We will just focus on the metric in the fiber directions, which according to (4.6) is given by

$$f''(s)\frac{\sqrt{-1}dw \wedge d\bar{w}}{|w|^2}.$$

Let us write $r = |w|$, so $s = 2\log r$. The condition that this metric extends across $w = 0$ is that f'' has the form

$$f''(s) = c_2r^2 + c_3r^3 + c_4r^4 + \cdots .$$

Then, since $d/ds = \frac{r}{2}d/dr$, we have

$$f'''(s) = c_2r^2 + \frac{3}{2}c_3r^3 + 2c_4r^4 + \cdots ,$$

and since $f''(s) = \varphi(\tau)$ and $f'''(s) = \varphi'(\tau)\varphi(\tau)$, we have

$$\varphi'(\tau) = 1 + O(r).$$

In particular if the range of τ is an interval (a, b), then

$$\lim_{\tau\to a}\varphi(\tau) = 0, \qquad \lim_{\tau\to a}\varphi'(\tau) = 1.$$

A similar computation can be done as $w \to \infty$ by changing coordinates to w^{-1}, showing that

$$\lim_{\tau\to b}\varphi(\tau) = 0, \qquad \lim_{\tau\to b}\varphi'(\tau) = -1.$$

Note also that by (4.8) the metric will be positive definite as long as $(1+\tau)$ and $\varphi(\tau)$ are positive on $[a, b]$. For simplicity we can take the interval $[0, m]$ for some $m > 0$. The value of m determines the Kähler class of the resulting metric. Viewing X as a $\mathbf{CP}^1$-bundle over Σ, the space $H^2(X, \mathbf{R})$ is generated by Poincaré duals of a fiber C and the infinity section S_∞, which is the image of the subbundle $L \oplus \{0\} \subset L \oplus \mathcal{O}$ under the projection map to the projectivization $X = \mathbf{P}(L \oplus \mathcal{O})$. We have the following intersection formulas:

$$C \cdot C = 0, \quad S_\infty \cdot S_\infty = 1, \quad C \cdot S_\infty = 1.$$

The Kähler class of the metric can then be determined by computing the areas of C and S_∞. The area of C is given by

$$\int_{\mathbf{C}\setminus\{0\}} f''(s) \frac{\sqrt{-1} dw \wedge d\bar{w}}{|w|^2} = 2\pi(\lim_{s\to\infty} f'(s) - \lim_{s\to-\infty} f'(s)) = 2\pi m,$$

while the area of the infinity section S_∞ is

$$(1+m)\int_\Sigma \omega_\Sigma = 2\pi(1+m).$$

It follows that if we denote by $\mathcal{L}_m$ the Poincaré dual to the Kähler class of ω, then

$$\mathcal{L}_m = 2\pi(C + mS_\infty).$$

The final thing to check is when the metric is extremal, i.e. when $\mathrm{grad}^{1,0} S(\tau)$ is holomorphic. We can compute that

$$\mathrm{grad}^{1,0} S(\tau) = S'(\tau) w \frac{\partial}{\partial w},$$

which is a holomorphic vector field if and only if $S'(\tau)$ is constant. So ω is extremal if and only if $S''(\tau) = 0$.

The end result is the following theorem, which follows from the more general results in Hwang-Singer [**65**].

Theorem 4.31. *Suppose that $\varphi : [0, m] \to \mathbf{R}$ is a smooth function which is positive on $(0, m)$ and satisfies the boundary conditions*

$$\varphi(0) = \varphi(m) = 0, \quad \varphi'(0) = 1, \quad \varphi'(m) = -1. \tag{4.10}$$

Then by the above construction we obtain a metric on X in the Kähler class Poincaré dual to $\mathcal{L}_m = 2\pi(C + mS_\infty)$ whose scalar curvature is given by

$$S(\tau) = -\frac{2}{1+\tau} - \frac{1}{1+\tau}[(1+\tau)\varphi]''.$$

The metric is extremal if and only if $S''(\tau) = 0$.

We can now construct extremal metrics by solving the ODE $S''(\tau) = 0$ for $\varphi : [0, m] \to \mathbf{R}$ satisfying the boundary conditions (4.10). The equation to be solved is

$$\frac{-1}{1+\tau}\Big(2 + \big[(1+\tau)\varphi\big]''\Big) = A + B\tau,$$

for some A, B. This equation can easily be integrated using the boundary conditions, and we obtain

$$\varphi(\tau) = \frac{\tau(m-\tau)}{m(m^2+6m+6)(1+\tau)}\big[\tau^2(2m+2) + \tau(-m^2+4m+6) \\ + m^2+6m+6\big].$$

This will only give rise to a metric if $\varphi(\tau) > 0$ for all $\tau \in (0, m)$. This happens only if $m < k_1$, where $k_1 \approx 18.889$ is the positive root of $m^4 - 16m^3 - 52m^2 - 48m - 12$. We have therefore constructed extremal metrics with non-constant scalar curvature on the $\mathbf{CP}^1$-bundle X in the Kähler classes Poincaré dual to $\mathcal{L}_m$ for $m < k_1$.

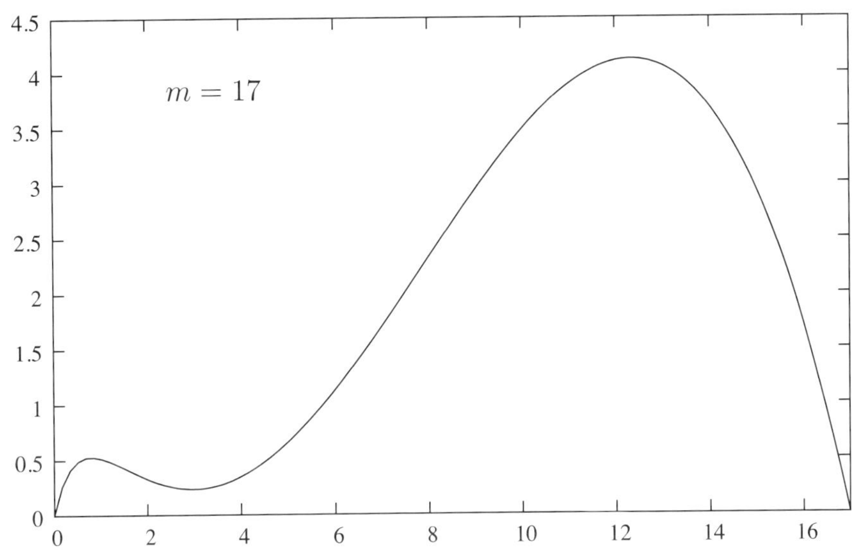

Figure 4.1. The momentum profile for the extremal metric when $m = 17$

It is interesting to see what happens as $m \to k_1$. At $m = k_1$ the solution $\varphi(\tau)$ acquires a zero in $(0, m)$ (Figure 4.1 shows the graph of φ when $m = 17$). Geometrically this corresponds to the fiber metrics degenerating in such a way that the diameter becomes unbounded but the area remains bounded. In other words the fibers break up into two pieces, each with an end asymptotic to a hyperbolic cusp. We will see in Section 6.5 that X is not relatively K-stable when $m \geqslant k_1$, and it follows that it does not admit an extremal metric for these Kähler classes.

Exercise 4.32. Show that the blow-up $\mathrm{Bl}_p\mathbf{CP}^2$ of the projective plane in one point admits an extremal metric in every Kähler class. Use the fact that we can write $\mathrm{Bl}_p\mathbf{CP}^2 = \mathbf{P}(\mathcal{O}(1) \oplus \mathcal{O})$ as a $\mathbf{CP}^1$-bundle over $\mathbf{CP}^1$, and so we can use the method above. An alternative approach is to exploit the fact that $\mathrm{Bl}_p\mathbf{CP}^2$ is a toric manifold and use the calculations in the next section. This is an example due to Calabi [**21**].

4.5. Toric manifolds

Toric manifolds are a fertile testing ground for many ideas in algebraic and symplectic geometry, and it turns out that the study of extremal metrics on them is also very fruitful. The basic Kähler geometry of toric manifolds was worked out by Guillemin [**61**], and the study of extremal metrics on them was initiated by Abreu [**1**]. This was then taken considerably further by a sequence of works by Donaldson [**44**], [**45**], [**48**], [**49**], culminating in a general existence result for cscK metrics on K-stable toric surfaces, with a further extension to the extremal case by Chen-Li-Sheng [**28**]. In this section we will discuss Kähler metrics on toric manifolds and Abreu's formula for their scalar curvature.

There are many different descriptions of toric manifolds from the point of view of algebraic geometry and symplectic geometry. In this section, the main point for us is that an n-dimensional toric manifold M contains a dense open set biholomorphic to $\mathbf{T}_{\mathbf{C}} = (\mathbf{C}^*)^n$, and the action of this complex torus on itself extends in a smooth way to an action on all of M.

Similarly to what we did in Section 4.4 we will be interested in Kähler metrics on M which on $(\mathbf{C}^*)^n$ can be written as

$$\omega = \sqrt{-1}\partial\overline{\partial} f(x), \tag{4.11}$$

where $x \in \mathbf{R}^n$ has components $x^i = \log|z^i|^2$ for $(z_1, \ldots, z_n) \in (\mathbf{C}^*)^n$. In terms of local complex coordinates $w^i = \log z^i$, we can compute

$$\sqrt{-1}\partial\overline{\partial} f(x) = \sqrt{-1}\frac{\partial^2 f}{\partial x^i \partial x^j} dw^i \wedge d\overline{w}^j, \tag{4.12}$$

so ω is a Kähler metric whenever $f : \mathbf{R}^n \to \mathbf{R}$ is strictly convex. The function f needs to satisfy certain conditions at infinity for this metric to extend to M, but as in the previous section, it is useful to first take a Legendre transform. We thus introduce the variable

$$y = \nabla f(x)$$

and define the function u by

$$f(x) + u(y) = x \cdot y. \tag{4.13}$$

The function u is called the *symplectic potential* of the metric, and it is a convex function on a subset $P \subset \mathbf{R}^n$ given by the range of ∇f. The set P turns out to be the interior of a polytope, and ∇f is a moment map for the action of $(S^1)^n$; see Example 5.7 for more details. For this reason P is called the *moment polytope*. The scalar curvature of ω has a particularly nice form in terms of u due to Abreu [**1**].

Proposition 4.33. *The scalar curvature of ω as a function of the y^i is given by*

$$S(\omega) = -\sum_{j,k} \frac{u^{jk}}{\partial y^j \partial y^k},$$

where u^{jk} is the inverse of the Hessian of u.

Proof. From expression (4.12), the Ricci form of ω in terms of the w^i is given by

$$\begin{aligned} R_{j\bar{k}} &= -\partial_j \partial_{\bar{k}} \log \det(f_{ab}) \\ &= -\frac{\partial^2}{\partial x^j \partial x^k} \log \det(f_{ab}), \end{aligned}$$

where f_{ab} denotes the Hessian of f in the x^i variables. From the definition of the Legendre transform we have

$$\frac{\partial}{\partial x^j} = \frac{\partial y^p}{\partial x^j} \frac{\partial}{\partial y^p}$$

and also

$$f_{ab}(x) = u^{ab}(y),$$

where u^{ab} is the inverse of the Hessian of u in the y^i variables. It then follows that

$$\begin{aligned} S(\omega) &= f^{jk} \frac{\partial y^p}{\partial x^j} \frac{\partial}{\partial y^p} \left(\frac{\partial y^q}{\partial x^k} u^{ab} \frac{\partial}{\partial y^q} u_{ab} \right) \\ &= u_{jk} u^{pj} \frac{\partial}{\partial y^p} \left(u^{kq} u^{ab} \frac{\partial}{\partial y^a} u_{qb} \right) \\ &= -\frac{\partial}{\partial y^k} \frac{\partial}{\partial y^a} u^{ak}, \end{aligned}$$

where we sum over repeated indices. This formula is what we wanted to prove. □

In order to decide when ω is an extremal metric, we need to know when $\mathrm{grad}^{1,0} h$ is holomorphic for a function h of the variables y^i.

Lemma 4.34. *The vector field $\mathrm{grad}^{1,0} h(y)$ is holomorphic if and only if h is an affine linear function of y.*

Proof. Suppose that $v^j = g^{j\bar{k}}\partial_{\bar{k}}h$ in terms of the variables w^i. Then

$$\begin{aligned}\partial_{\bar{p}}v^j &= \partial_{\bar{p}}(g^{j\bar{k}}\partial_{\bar{k}}h)\\ &= \frac{\partial}{\partial x^p}\left(f^{jk}\frac{\partial h}{\partial x^k}\right)\\ &= u^{pq}\frac{\partial}{\partial y^q}\left(u_{jk}u^{kl}\frac{\partial h}{\partial y^l}\right)\\ &= u^{pq}\frac{\partial^2 h}{\partial y^q \partial y^j}.\end{aligned}$$

It follows that v^j is holomorphic if and only if h is affine linear. □

We will now briefly discuss the question of when the metric ω extends to M from $\mathbf{T_C}$. The points in $M \setminus \mathbf{T_C}$ have non-trivial stabilizer, and we can classify them according to the dimension of the stabilizer. Let us focus on a fixed point $p \in M$ of the torus action, where the stabilizer is the whole n-dimensional torus. We can choose local coordinates z^i centered at p such that the action of $\mathbf{T_C}$ is given by componentwise multiplication. Suppose for simplicity that ω is given by

$$\omega = \sqrt{-1}\sum_j dz^j \wedge d\overline{z^j}$$

in a neighborhood of p. In terms of the w^i and x^i we have

$$\begin{aligned}\omega &= \sqrt{-1}\sum_j e^{2\mathrm{Re}\, w^j} dw^j \wedge d\overline{w^j}\\ &= \sqrt{-1}\sum_j e^{x^j} dw^j \wedge d\overline{w^j},\end{aligned}$$

and so up to the addition of an affine linear function we have

$$f(x) = e^{x^1} + \cdots + e^{x^n}.$$

Taking the Legendre transform we obtain

$$u(y) = \sum_j \left(y^j \ln y^j - y^j\right), \tag{4.14}$$

where $y^j = e^{x^j}$. The point p corresponds to $y^j = 0$, and the domain of u is a neighborhood of the origin in the positive orthant. Modifying f by an affine linear function amounts to a translation in the y variables, while choosing a different integral basis for the torus transforms the x and y variables by an element of $SL(n, \mathbf{Z})$.

More generally we can work at a point $p \in M$ where the stabilizer is a k-dimensional torus. We can then choose coordinates z^i in which

$$p = (\overbrace{0, \ldots, 0}^{k}, \overbrace{1, \ldots, 1}^{n-k}),$$

and the torus action is still given by componentwise multiplication. Still using the Euclidean metric as above, if we take the Legendre transform, then we obtain the function u as in (4.14), but now p corresponds to the point $y = (0, \ldots, 0, 1, \ldots, 1)$. The domain of u will be a neighborhood of this point inside the positive orthant and once again different choices of coordinates, and modifying f by a linear function amounts to an $SL(n, \mathbf{Z})$ transformation and a translation.

From expression (4.12) for the metric in terms of f we can see that if g is an $(S^1)^n$-invariant function on $\mathbf{T_C}$ which extends smoothly to M, then in terms of the x variables we have $\nabla g \to 0$ at infinity. It follows that the image of ∇f, i.e. the domain of y, does not change when we modify our metric ω by a globally defined Kähler potential. This means that the information that we obtained above about the possible domains of y near points in $M \setminus \mathbf{T_C}$ applies even if we do not use the Euclidean metric. Piecing this information together we see that the domain of y is the interior of a convex polytope $P \subset \mathbf{R}^n$ and a neighborhood of each vertex of P is equivalent to a neighborhood of the origin in the positive orthant, under a translation and the action of $SL(n, \mathbf{Z})$. This means that P satisfies the following Delzant condition.

Definition 4.35. Let $P \subset \mathbf{R}^n$ be a convex polytope defined by a set of inequalities

$$l_i(y) \geqslant c_i, \tag{4.15}$$

where the l_i are linear functions with coprime integral coefficients and $c_i \in \mathbf{R}$. We say that P satisfies the *Delzant condition* if n faces meet at each vertex $p \in P$, given by equations

$$l_1(y) = c_1, \ldots, l_n(y) = c_n,$$

where the l_i generate the dual space $(\mathbf{Z}^n)^*$ over $\mathbf{Z}$.

Based on the observations above, one has the following result (see Abreu [**1**] for details of the proof).

Theorem 4.36. *Let (M, ω) be a toric Kähler manifold with moment polytope $P \subset \mathbf{R}^n$ defined by inequalities (4.15). Then every $(S^1)^n$-invariant Kähler metric in the class $[\omega]$ has a symplectic potential $u : P \to \mathbf{R}$ of the* *form*

$$u = \sum_j (l_i(y) - c_i)\ln(l_i(y) - c_i) + v, \tag{4.16}$$

where v is smooth up to the boundary of P, while u is strictly convex on the interior of P and its restriction to each facet of P is strictly convex on the interior of that facet.

Let us denote by $\mathcal{S}$ the set of functions $u : P \to \mathbf{R}$ of the form (4.16) satisfying the conditions in the theorem. In summary we see that to find a torus invariant extremal metric on M in a given Kähler class, we need to find $u \in \mathcal{S}$ such that $S(u)$ is affine linear, where

$$S(u) = -\sum_{j,k} \frac{\partial^2 u^{jk}}{\partial y^j \partial y^k}.$$

To conclude this section we will examine what the Futaki invariant, the Mabuchi functional, and geodesics correspond to in terms of symplectic potentials. The following basic integration by parts formula can be found in Donaldson [**44**].

Lemma 4.37. *Suppose that $u \in \mathcal{S}$, and let $g : P \to \mathbf{R}$ be a continuous convex function that is smooth on the interior of P. Then*

$$\int_P u^{jk} g_{jk}\, d\mu = \int_{\partial P} g\, d\sigma - \int_P gS(u)\, d\mu,$$

where $d\mu$ is the Lebesgue measure on P, while $d\sigma$ is a positive measure on the boundary ∂P normalized so that on a face defined by $l_i(y) = c_i$ as in (4.15) *we have $d\sigma \wedge dl_i = \pm d\mu$. In addition u^{jk} and g_{jk} are the inverse Hessian of u and the Hessian of g, respectively.*

Note that in terms of the variables x^i, θ^i, where $w_i = \frac{1}{2}x^i + \sqrt{-1}\theta^i$, the volume form of the metric ω in (4.11) is

$$\frac{\omega^n}{n!} = \det(f_{jk}) dx^1 \wedge d\theta^1 \wedge \cdots \wedge dx^n \wedge d\theta^n,$$

which after transforming to the y^i variables becomes

$$\frac{\omega^n}{n!} = dy^1 \wedge d\theta^1 \wedge \cdots \wedge dy^n \wedge d\theta^n.$$

It follows that the integral of a function $g(y)$ on (M, ω) is simply the integral of g on P up to a factor of $(2\pi)^n$. Applied to $g = 1$, Lemma 4.37 then implies that the average of the scalar curvature $S(u)$ is the constant

$$a = \frac{\int_P S(u)\, d\mu}{\int_P d\mu} = \frac{\mathrm{Vol}(\partial P, d\sigma)}{\mathrm{Vol}(P, d\mu)}.$$

We have already seen that the holomorphy potentials on (M,ω) correspond to affine linear functions h on P. Formula (4.1) defining the Futaki invariant then gives

$$\begin{aligned}(2\pi)^{-n}F(h) &= \int_P h(S(u)-a)\,d\mu \\ &= \int_{\partial P} h\,d\sigma - a\int_P h\,d\mu.\end{aligned}$$

The Futaki invariant vanishes for all vector fields in $\mathfrak{h}$ when $F(h)=0$ for all affine linear functions h. This is equivalent to saying that the center of mass of $(P, a\,d\mu)$ equals the center of mass of $(\partial P, d\sigma)$.

Note that Lemma 4.37 implies a simple necessary condition for the existence of a symplectic potential $u\in\mathcal{S}$ with constant scalar curvature. Indeed, if $S(u)$ is constant, then necessarily $S(u)=a$, and so for every convex smooth function g on P which is not affine linear, we have

$$\int_{\partial P} g\,d\sigma - a\int_P g\,d\mu = \int_P u^{jk}g_{jk}\,d\mu > 0.$$

More generally, if $S(u)=A$ for an affine linear function A, then the same argument implies that

$$\int_{\partial P} g\,d\sigma - \int_P Ag\,d\mu > 0, \tag{4.17}$$

for all non-affine linear convex functions g. We will relate this condition to stability in Section 6.7.

Let us turn now to the Mabuchi functional. We have the following from [**44**].

Proposition 4.38. *With a suitable normalization by adding a constant, the Mabuchi functional evaluated at $u\in\mathcal{S}$ is given, up to a factor of $(2\pi)^n$, by*

$$\mathcal{M}(u) = -\int_P \log\det(u_{ab})\,d\mu + \int_{\partial P} u\,d\sigma - a\int_P u\,d\mu. \tag{4.18}$$

Proof. To see this, it is enough to check that the variation of the functional given by (4.18) matches up with the definition of the Mabuchi functional in (4.2). If $u_t = u+tv\in\mathcal{S}$, then we have

$$\begin{aligned}\left.\frac{d}{dt}\right|_{t=0}\mathcal{M}(u_t) &= -\int_P u^{ab}v_{ab}\,d\mu + \int_{\partial P} v\,d\sigma - a\int_P v\,d\mu \\ &= \int_P v(S(u)-a)\,d\mu.\end{aligned} \tag{4.19}$$

At the same time, if the $f_t(x)$ are the Legendre transforms of $u_t(y)$, then differentiating formula (4.13) we get

$$\frac{d}{dt}\Big|_{t=0} f_t(x) = -\frac{d}{dt}\Big|_{t=0} u_t(y).$$

It follows that the variation of $\mathcal{M}$ in (4.18) matches up with the variation of the Mabuchi functional in (4.2). □

We now turn to geodesics of toric Kähler metrics.

Proposition 4.39. *A family $u_t \in \mathcal{S}$ of symplectic potentials corresponds to a (constant speed) geodesic of Kähler metrics if and only if* $\frac{d^2}{dt^2}u_t = 0$.

Proof. Rather than rewriting the geodesic equation from Proposition 4.25 in terms of symplectic potentials, we will derive the equation again from the energy of a path. The key point is that in terms of the y variables the volume form is fixed. The energy of the path u_t, for $t \in [0,1]$, say, is

$$E(u_t) = \int_0^1 \int_P \dot{u}_t^2 \, d\mu \, dt,$$

so given a variation $u_t + \varepsilon v_t$ with $v_0 = v_1 = 0$, we have

$$\begin{aligned} \frac{d}{d\varepsilon}\Big|_{\varepsilon=0} E(u_t + \varepsilon v_t) &= \int_0^1 \int_P 2\dot{u}_t \dot{v}_t \, d\mu \, dt \\ &= \int_0^1 \int_P -2\ddot{u}_t v_t \, d\mu \, dt. \end{aligned}$$

It follows that critical points of E satisfy $\ddot{u}_t = 0$. □

In particular any two symplectic potentials u_0, u_1 can be joined by a smooth geodesic using linear interpolation. The following uniqueness result is a simple consequence of this.

Proposition 4.40. *Suppose that $S(u_0) = S(u_1) = a$ for symplectic potentials $u_0, u_1 \in \mathcal{S}$. Then $u_0 - u_1$ is an affine linear function.*

Proof. Consider the geodesic $u_t = u_0 + tv$, where $v = u_1 - u_0$. We have

$$\frac{d}{dt}\mathcal{M}(u_t) = \int_P v(S(u_t) - a) \, d\mu,$$

so by our assumption the derivative of $\mathcal{M}(u_t)$ vanishes for $t = 0$ and $t = 1$. At the same time, from Proposition 4.27 we know that $\mathcal{M}(u_t)$ is convex. It follows that $\mathcal{M}(u_t)$ must be a constant and so from equation (4.3) we see that $v = u_1 - u_0$ must be a holomorphy potential, i.e. an affine linear function. □

We conclude this section with an example.

Example 4.41. Let $M = \mathbf{CP}^2$ with homogeneous coordinates $[Z^0 : Z^1 : Z^2]$. The points $[1 : z^1 : z^2]$ for $(z^1, z^2) \in (\mathbf{C}^*)$ define a dense complex torus, and the natural multiplication action extends as

$$(z^1, z^2) \cdot [Z^0 : Z^1 : Z^2] = [Z^0 : z_1 Z^1 : z_2 Z^2].$$

This action has three fixed points, $[1 : 0 : 0]$, $[0 : 1 : 0]$, and $[0 : 0 : 1]$, corresponding to vertices p_0, p_1, p_2 of the moment polytope P. We can work out what the moment polytope looks like near these fixed points as follows. Near the point $[1 : 0 : 0]$ the action is standard, relative to the coordinates

$$z^1 = \frac{Z^1}{Z^0}, \qquad z^2 = \frac{Z^2}{Z^0},$$

so a neighborhood of p_0 is a translation of a neighborhood of the origin in the first quadrant. Near $[0 : 1 : 0]$ and $[0 : 0 : 1]$ we have the following coordinates:

$$\text{near } [0:1:0]: \frac{Z^2}{Z^1} = \frac{z^2}{z^1}, \frac{Z^0}{Z^1} = \frac{1}{z^1},$$
$$\text{near } [0:0:1]: \frac{Z^0}{Z^2} = \frac{1}{z^2}, \frac{Z^1}{Z^2} = \frac{z^1}{z^2}.$$

These are related to the basis z^1, z^2 of the torus by the $SL(2, \mathbf{Z})$ matrices

$$\begin{pmatrix} -1 & -1 \\ 1 & 0 \end{pmatrix}, \quad \begin{pmatrix} 0 & 1 \\ -1 & -1 \end{pmatrix},$$

which transform the "standard corner" spanned by the vectors $(1, 0), (0, 1)$ into corners spanned by $(-1, 1), (-1, 0)$ and $(0, -1), (1, -1)$. It follows that P is a right triangle with two equal sides parallel to the x and y axes. The size of the triangle is determined by the Kähler class, while its location in the plane is determined by the choice of normalization for the Kähler potential on $\mathbf{T_C}$, or equivalently a choice of moment map for the $(S^1)^n$ action.

Suppose that our Kähler class is chosen in such a way that P has vertices $(0, 0)$, $(1, 0)$, and $(0, 1)$. Then a symplectic potential on P is given by

$$u = x \ln x + y \ln y + (1 - x - y) \ln(1 - x - y).$$

A straightforward although tedious calculation shows that $S(u) = 6$, so u corresponds to a cscK metric. In fact it is the Fubini-Study metric on $\mathbf{CP}^2$.

In general it is a difficult problem to find symplectic potentials giving rise to extremal metrics, and a complete existence theory has so far only been worked out in the 2-dimensional case. We will discuss this briefly in Section 6.7, where we study the algebro-geometric side of the problem.

Chapter 5

Moment Maps and Geometric Invariant Theory

The main result in this chapter is the Kempf-Ness theorem, which relates quotient constructions in symplectic and algebraic geometry. When a compact group G acts by Hamiltonian isometries on a Kähler manifold M, then there is a moment map $\mu : M \to \mathfrak{g}^*$ to the dual of the Lie algebra. The Kempf-Ness theorem characterizes those orbits $G^c \cdot p$ of the complexified group which contain zeros of the moment map in terms of algebro-geometric stability. In symplectic geometry quotients are constructed by taking the G-quotient of $\mu^{-1}(0)$, while in algebraic geometry the quotient parametrizes the stable orbits and the Kempf-Ness theorem implies that the two quotient constructions give the same result.

Stability can be tested using the Hilbert-Mumford criterion discussed in Section 5.3, and this will be a motivation for the definition of K-stability in the next chapter. For a much more thorough treatment of this subject see Mumford-Fogarty-Kirwan [**85**], or see Thomas [**109**] for an exposition with extremal metrics in mind.

5.1. Moment maps

Let (M, ω) be a compact Kähler manifold with Kähler metric g. We could work more generally with a symplectic manifold, but it is convenient to have the Kähler structure. The Hamiltonian construction assigns a vector field

X_h on M to any smooth function $h : M \to \mathbf{R}$ satisfying

$$dh(Y) = -\omega(X_h, Y) = -\iota_{X_h}\omega(Y),$$

where ι_{X_h} is the contraction with X_h (contracting the first component). In terms of the metric g we have $dh(Y) = -g(JX_h, Y)$, so

$$X_h = J\mathrm{grad}\, h,$$

using the Riemannian gradient.

Lemma 5.1. *We have $L_{X_h}\omega = 0$, where L_{X_h} is the Lie derivative. In other words the one-parameter group of diffeomorphisms generated by X_h preserves the form ω.*

Proof. The Lie derivative satisfies the formula

$$L_{X_h}\omega = d(\iota_{X_h}\omega) + \iota_{X_h}d\omega.$$

This can be checked easily for 2-forms of the type $f\, dg \wedge dh$ and extended to arbitrary 2-forms by linearity. Since $d\omega = 0$, we have

$$L_{X_h}\omega = d(-dh) = 0,$$

using that $\iota_{X_h} = -dh$. □

Note that while X_h preserves ω, it does not preserve the metric g in general unless X_h is a real holomorphic vector field (i.e. $L_{X_h}J = 0$ for the complex structure J).

Suppose now that a connected Lie group G acts on M, preserving the form ω. The derivative of the action gives rise to a Lie algebra map

$$\rho : \mathfrak{g} \to \mathrm{Vect}(M),$$

where $\mathfrak{g}$ is the Lie algebra of G and $\mathrm{Vect}(M)$ is the space of vector fields on M. Roughly speaking, the action of G is called Hamiltonian if each of the vector fields in the image of ρ arises from the Hamiltonian construction.

Definition 5.2. The action of G on M is Hamiltonian if there exists a G-equivariant map

$$\mu : M \to \mathfrak{g}^*$$

to the dual of the Lie algebra of $\mathfrak{g}$ such that for any $\xi \in \mathfrak{g}$ the function $\langle \mu, \xi \rangle$ is a Hamiltonian function for the vector field $\rho(\xi)$:

$$d\langle \mu, \xi \rangle = -\omega(\rho(\xi), \cdot).$$

The action of G on $\mathfrak{g}^*$ is by the coadjoint action. The map μ is called a moment map for the action.

Equivalently, the action is Hamiltonian if there is a G-equivariant lift $m : \mathfrak{g} \to C^\infty(M)$ of the map ρ, where G acts on $\mathfrak{g}$ by the adjoint action. In the diagram below, Ham refers to the Hamiltonian construction:

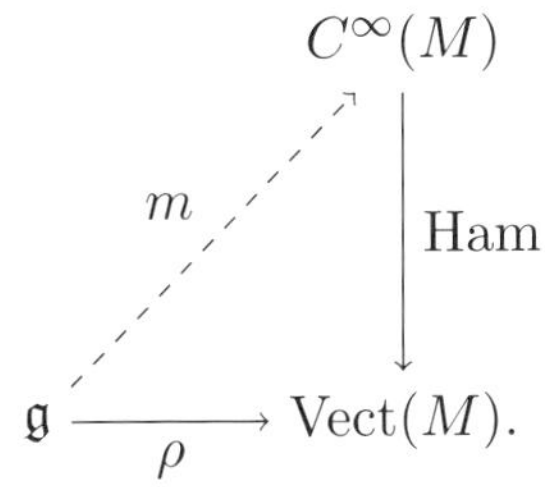

For any given Hamiltonian vector field X the possible lifts to $C^\infty(M)$ all differ by addition of constants. The G-equivariance requires a consistent choice of such lifts. In practice this is usually easily achieved by choosing a "natural" normalization for the Hamiltonian functions, for example requiring them to have average zero if M is compact. On the other hand there are cases when the G-equivariance cannot be achieved.

The moment map is important in constructing quotients of symplectic manifolds. In the above set-up, with a Hamiltonian action of G on M and a choice of moment map μ, the symplectic quotient is defined to be $\mu^{-1}(0)/G$. If the action of G on $\mu^{-1}(0)$ is free, then this quotient inherits a natural symplectic structure from M. If M is Kähler and the group G acts by isometries, then the quotient will inherit a Kähler structure. The basic idea is that at $x \in \mu^{-1}(0)$ the tangent space $T_x\mu^{-1}(0)$ is the kernel of $d\mu_x$, but from the definitions this is the orthogonal complement $(JT_xGx)^\perp$, where Gx is the G-orbit of x. We therefore have an identification

$$T_x\left(\mu^{-1}(0)/G\right) = (T_xGx \oplus JT_xGx)^\perp.$$

This is a complex subspace of T_xM, and the restrictions of the complex structure and the symplectic form define the Kähler structure on $\mu^{-1}(0)/G$. For more details on this see McDuff-Salamon [**84**].

Example 5.3. Consider the action $U(1) \curvearrowright \mathbf{C}$, by multiplication, and let $\omega = \sqrt{-1}dz \wedge d\bar{z} = 2dx \wedge dy$ be the standard Kähler form on $\mathbf{C}$. The action is generated by the vector field

$$X = x\frac{\partial}{\partial y} - y\frac{\partial}{\partial x},$$

so

$$\iota_X\omega = -2x\,dx - 2y\,dy.$$

So $h = x^2 + y^2$ satisfies $dh = -\iota_X\omega$. In other words,

$$\mu(z) = |z|^2$$

is a moment map for this action, after a suitable identification of $\mathfrak{u}(1)^*$ with $\mathbf{R}$. Other moment maps are given by $\mu(z) + c$ for any $c \in \mathbf{R}$.

Example 5.4. Generalizing the previous example, let $U(n) \curvearrowright \mathbf{C}^n$ be the standard action. For any given $A \in \mathfrak{u}(n)$ which generates a circle action, we can diagonalize A and apply the result of the previous example to each factor of $\mathbf{C}$. If A is diagonal with eigenvalues $\sqrt{-1}\lambda_i$ for $\lambda_i \in \mathbf{Z}$, then a Hamiltonian function for the vector field X_A generated by A is given by

$$z = (z_1, \ldots, z_n) \mapsto \lambda_1|z_1|^2 + \cdots + \lambda_n|z_n|^2 = -\sqrt{-1}\bar{z}^t A z.$$

The same formula then holds for any A. This means that a moment map for the action is given by

$$\begin{aligned} \mu : \mathbf{C}^n &\to \mathfrak{u}(n)^* \\ (z_1, \ldots, z_n) &\mapsto \sqrt{-1} z_i \bar{z}_j, \end{aligned}$$

where $\sqrt{-1}z_i\bar{z}_j$ defines a matrix in $\mathfrak{u}(n)$, and we identify $\mathfrak{u}(n)^* \simeq \mathfrak{u}(n)$ using the pairing $\langle A, B\rangle = -\mathrm{Tr}(AB)$.

Example 5.5. Consider now the action $U(n+1) \curvearrowright \mathbf{CP}^n$, which preserves the Fubini-Study form. $\mathbf{CP}^n$ is obtained from $\mathbf{C}^{n+1}$ as a symplectic quotient, with respect to the diagonal action of $U(1)$. More precisely, we choose the moment map

$$z \mapsto |z|^2 - 1$$

for this $U(1)$-action on $\mathbf{C}^{n+1}$. Then

$$\mathbf{CP}^n = (\{|z|^2 - 1 = 0\} \subset \mathbf{C}^{n+1})/U(1),$$

and the Fubini-Study form is the induced form on this quotient. The moment map on $\mathbf{CP}^n$ is therefore the moment map on $\mathbf{C}^{n+1}$ restricted to the subset where $|z|^2 = 1$. So we obtain the moment map

$$\begin{aligned} \mu : \mathbf{CP}^n &\to \mathfrak{u}(n+1)^* \\ [Z_0 : Z_1 : \cdots : Z_n] &\mapsto \frac{\sqrt{-1}Z_i\overline{Z}_j}{|Z|^2}. \end{aligned}$$

Here again $\mathfrak{u}(n+1)$ is identified with its dual using the pairing $-\mathrm{Tr}(AB)$. If we restrict the action to $SU(n+1)$, then the resulting moment map μ_{SU} is just the projection of μ onto $\mathfrak{su}(n+1)$, i.e.

$$\mu_{SU}([Z_0 : \cdots : Z_n]) = \frac{\sqrt{-1}Z_i\overline{Z}_j}{|Z|^2} - \frac{\sqrt{-1}}{n+1}\mathrm{Id},$$

where Id is the identity matrix. We can view $\mathbf{CP}^n$ as a coadjoint orbit in $\mathfrak{su}(n+1)^*$, and then μ_{SU} is the identity map.

Example 5.6. Consider the diagonal action $SU(2) \curvearrowright \mathrm{Sym}^n\mathbf{CP}^1$ on unordered n-tuples of points on $\mathbf{CP}^1$. We can identify $\mathfrak{su}(2)^*$ with $\mathbf{R}^3$, and $\mathbf{CP}^1$ with the unit sphere in $\mathbf{R}^3$ (as a coadjoint orbit). Under these identifications a moment map for the action is given by

$$\mu : \mathrm{Sym}^n\mathbf{CP}^1 \to \mathbf{R}^3$$
$$\mu(x_1, \ldots, x_n) = x_1 + \cdots + x_n.$$

This means that zeros of the moment map are given by n-tuples of points whose center of mass is the origin.

Example 5.7. Recall that in Section 4.5, for a strictly convex function $f : \mathbf{R}^n \to \mathbf{R}$ we defined a Kähler metric

$$\omega = \sqrt{-1}\sum_{j,k} \frac{\partial^2 f}{\partial x^j \partial x^k} dw^j \wedge d\overline{w}^k$$

on the complex torus $\mathbf{T}_{\mathbf{C}}$, where $w^j = \frac{1}{2}x^j + \sqrt{-1}\theta^j$. The $(S^1)^n$-action is generated by the vector fields ∂_{θ^j}, and we can compute

$$\iota_{\partial_{\theta^i}}\omega = -\sum_k \frac{\partial^2 f}{\partial x^i \partial x^k} dx^k.$$

It follows that a moment map for the $(S^1)^n$-action is given by the gradient map

$$\nabla f : \mathbf{R}^n \to \mathbf{R}^n.$$

In other words the domain of the Legendre transform of f is the image of the moment map. If, as in Section 4.5, the torus $\mathbf{T}_{\mathbf{C}}$ is a dense open subset of a compact Kähler manifold (M, ω) such that the torus action extends to all of M, then ∇f is the restriction to $\mathbf{T}_{\mathbf{C}}$ of a moment map $\mu : M \to \mathbf{R}^n$ for the $(S^1)^n$-action, where $\mathbf{R}^n$ is identified with the dual of the Lie algebra. Note that it is a general result due to Atiyah [**6**] and Guillemin-Sternberg [**62**] that the image of the moment map for any Hamiltonian torus action on a compact symplectic manifold is a convex polytope, which is the convex hull of the images of the fixed points of the action.

Exercise 5.8. Let M_n be the set of $n \times n$ complex matrices equipped with the Euclidean metric under the identification $M_n = \mathbf{C}^{n^2}$. The unitary matrices $U(n)$ act on M_n by conjugation, preserving this metric. I.e. $A \in U(n)$ acts by $M \mapsto A^{-1}MA$. Find a moment map

$$\mu : M_n \to \mathfrak{u}(n)^*$$

for this action (normalize it so that $\mu(0)$ is the zero matrix).

5.2. Geometric invariant theory (GIT)

Suppose that $M \subset \mathbf{CP}^n$ is a projective variety (see below for definitions) and a complex Lie group $G \subset GL(n+1, \mathbf{C})$ acts on M by biholomorphisms. More invariantly we could take a compact complex manifold M together with an ample line bundle L on it and an action of G on M together with a lifting of the action to L. The Kodaira embedding theorem implies, however, that up to replacing L by a power, this is the same as the more concrete situation above.

Geometric invariant theory gives a way of constructing a quotient M/G which is also a projective variety. The basic idea is that M/G should be characterized by the requirement that

$$\text{“functions on } M/G\text{”} = \text{“}G\text{-invariant functions on } M\text{”}.$$

Example 5.9. Before giving more precise definitions, let us look at a simple example that illustrates some of the ideas, although it does not fit precisely into the framework that we are considering since here we are working with affine varieties instead of projective ones. Suppose that $\mathbf{C}^*$ acts on $\mathbf{C}^2$ with the action

$$\lambda \cdot (x, y) = (\lambda x, \lambda^{-1} y).$$

There are three types of orbits:

(i) $xy = t$ for $t \neq 0$. These are closed 1-dimensional orbits.

(ii) $x = y = 0$. This is a closed 0-dimensional orbit.

(iii) $x = 0, y \neq 0$, or $x \neq 0, y = 0$. These are two 1-dimensional orbits whose closures contain the origin.

The orbit space is not Hausdorff because the closure of the orbits of type (iii) contain the orbit (ii). However if we discard the non-closed orbits (iii), then the remaining orbits are parametrized by $\mathbf{C}$. In terms of functions, the $\mathbf{C}^*$-invariant functions on $\mathbf{C}^2$ are

$$\mathbf{C}[x, y]^{\mathbf{C}^*} = \mathbf{C}[xy] \cong \mathbf{C}[t],$$

and so the space of functions of the quotient should be $\mathbf{C}[t]$. Therefore the quotient $\mathbf{C}^2/\mathbf{C}^*$ from this point of view should also be $\mathbf{C}$. Looking ahead a little, this example also illustrates the relationship with the moment map very clearly. The action of the maximal compact subgroup $U(1) \subset \mathbf{C}^*$ on $\mathbf{C}^2$ is Hamiltonian with respect to the standard symplectic form, and a moment map is given by (see Example 5.4)

$$\mu(x, y) = |x|^2 - |y|^2.$$

The symplectic quotient $\mu^{-1}(0)/U(1)$ also equals $\mathbf{C}$, and $\mu^{-1}(0)$ intersects each closed orbit of $\mathbf{C}^*$ in a $U(1)$-orbit. See Figure 5.1.

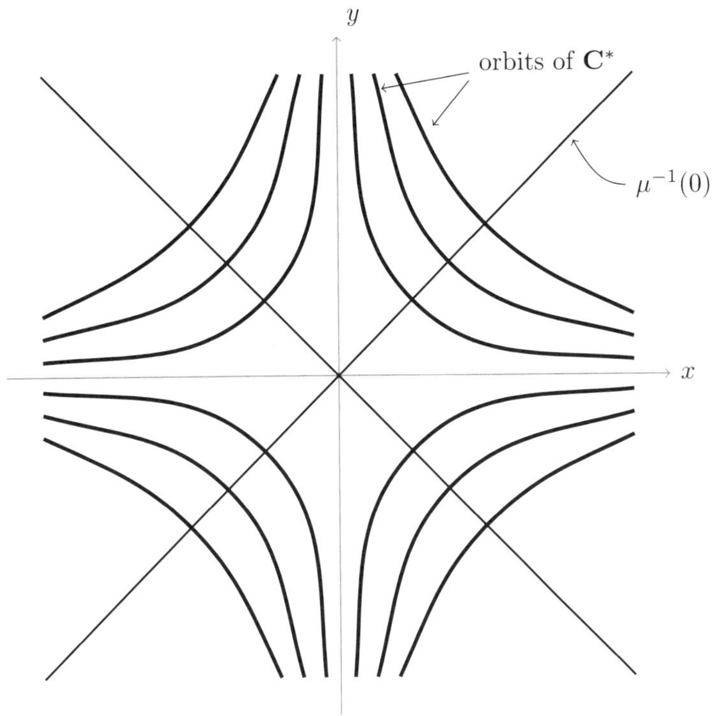

Figure 5.1. The orbits of $\mathbf{C}^* \curvearrowright \mathbf{C}^2$ and zeros of μ.

In order to define the GIT quotient in general, we need a quick review of some basic ideas in algebraic geometry.

Definition 5.10. (a) A projective variety $X \subset \mathbf{CP}^n$ is the zero set of a collection of homogeneous polynomials $f_1, \ldots, f_k$, which is irreducible, i.e. it cannot be written as a non-trivial union of two such zero sets.

(b) The homogeneous coordinate ring of X is the graded ring (graded by degree)

$$R(X) = \mathbf{C}[x_0, \ldots, x_n]/I,$$

where I is the ideal generated by the homogeneous polynomials vanishing on X. Since X is irreducible, the ideal I is prime (i.e. if $fg \in I$, then $f \in I$ or $g \in I$). Equivalently the ring $R(X)$ has no zero-divisors.

Conversely any homogeneous prime ideal $I \subset \mathbf{C}[x_0, \ldots, x_n]$ gives rise to a projective variety as long as $I \neq (x_0, \ldots, x_n)$ (in which case the zero set would be empty). The Nullstellensatz in commutative algebra implies that there is a one-to-one correspondence

$$\left\{\begin{array}{c}\text{homogeneous prime ideals}\\ \text{in } \mathbf{C}[x_0,\dots,x_n]\\ \text{except } (x_0,\dots,x_n)\end{array}\right\} \longleftrightarrow \left\{\begin{array}{c}\text{projective subvarieties}\\ \text{of } \mathbf{CP}^n\end{array}\right\}.$$

It is often convenient to work on the level of functions and forget the way that our variety is embedded in projective space. The correspondence in this case is

$$\left\{\begin{array}{c}\text{finitely generated graded } \mathbf{C}\text{-algebras,}\\ \text{generated in degree 1,}\\ \text{without zero-divisors}\end{array}\right\} \longleftrightarrow \left\{\begin{array}{c}\text{projective}\\ \text{varieties}\end{array}\right\}, \tag{5.1}$$

although to make this correspondence one-to-one, we would have to define equivalence relations on both sets. In this correspondence a projective variety is mapped to its homogeneous coordinate ring. For the converse direction, if R is a finitely generated graded $\mathbf{C}$-algebra, generated in degree 1, then there is a surjective grading-preserving map

$$\mathbf{C}[x_0,\dots,x_n] \to R$$

mapping each x_i to a degree 1 generator. If I is the kernel of this map, then

$$\mathbf{C}[x_0,\dots,x_n]/I \cong R,$$

and I is prime since R has no zero-divisors. The vanishing set of the homogeneous elements in I is the projective variety $X \subset \mathbf{CP}^n$ corresponding to R. Let us call this projective variety $\operatorname{Proj}(R)$.

Remark 5.11. In the theory of schemes, the above correspondences are extended by allowing arbitrary homogeneous ideals as opposed to just prime ideals, and correspondingly, arbitrary finitely generated $\mathbf{C}$-algebras, not just those without zero-divisors (even more generally one does not need to work over a field, but we do not need this fact). We will see later how these schemes arise naturally and how we can think of them geometrically.

With this background we can proceed to define the GIT quotient.

Definition 5.12. A complex Lie group G is reductive if it is the complexification of a maximal compact subgroup $K \subset G$. For example $GL(n,\mathbf{C})$ is reductive with maximal compact subgroup $U(n)$. In particular the group $\mathbf{C}^*$ is reductive with maximal compact subgroup $U(1)$. On the other hand the additive group $\mathbf{C}$ is not reductive since it has no non-trivial compact subgroups at all.

Suppose that a complex reductive group G acts on a projective variety $X \subset \mathbf{CP}^n$ and the action is induced by a representation

$$G \to GL(n+1,\mathbf{C}).$$

Through the dual action on $\mathbf{C}[x_0, \ldots, x_n]$, this induces an action of G on the homogeneous coordinate ring $R(X)$. Using that G is a reductive group, one can show that the ring of invariants $R(X)^G$ is finitely generated. Let us write

$$R(X)^G = \bigoplus_{k \geqslant 0} R(X)^G_k,$$

where $R(X)^G_k$ is the degree k piece. To get a projective variety, we would like to replace $R(X)^G$ with a $\mathbf{C}$-algebra which is generated in degree 1. For this, one shows that there is a $d > 0$ such that the subalgebra

$$\widetilde{R}(X)^G = \bigoplus_{k \geqslant 0} R(X)^G_{kd}$$

is generated by elements in $R(X)^G_d$. Changing the grading so that $R(X)^G_{kd}$ is the degree k piece in $\widetilde{R}(X)^G$, we obtain a $\mathbf{C}$-algebra generated in degree 1, and we define the GIT quotient to be

$$X /\!\!/ G = \operatorname{Proj} \widetilde{R}(X)^G.$$

Since $\widetilde{R}(X)^G$ is a subalgebra of $R(X)$, it has no zero-divisors, and so $X /\!\!/ G$ is a projective variety.

While this definition is very simple (at least once the correspondence (5.1) has been established), it is unclear at this point what the quotient $X /\!\!/ G$ represents geometrically. To understand this, let us choose degree 1 generators $f_0, \ldots, f_k$ of $\widetilde{R}(X)^G$ and look at the map

$$\begin{aligned} q : X &\dashrightarrow \mathbf{CP}^k \\ p &\mapsto [f_0(p), \ldots, f_k(p)], \end{aligned}$$

which is only defined at points $p \in X$ at which there is at least one non-vanishing G-invariant function in R. Then the image of q is $X /\!\!/ G$, and q is the quotient map. The main points are therefore the following:

(i) The quotient $X /\!\!/ G$ parametrizes orbits on which there is at least one non-vanishing G-invariant function in R.

(ii) The quotient map $q : X \dashrightarrow X /\!\!/ G$ identifies any two orbits which cannot be distinguished by G-invariant functions in R.

This motivates the following definitions.

Definition 5.13. The set of semistable points $X^{ss} \subset X$ is defined by

$$X^{ss} = \left\{ p \in X \;\middle|\; \begin{array}{c} \text{there exists a non-constant homogeneous } f \in R(X)^G \\ \text{such that } f(p) \neq 0 \end{array} \right\}.$$

The set of stable points $X^s \subset X^{ss}$ is defined by

$$X^s = \left\{ p \in X^{ss} \;\middle|\; \begin{array}{c} \text{the stabilizer of } p \text{ in } G \text{ is finite} \\ \text{and the orbit } G \cdot p \text{ is closed in } X^{ss} \end{array} \right\}.$$

Both X^s, X^{ss} are open subsets of X. The GIT quotient $X /\!\!/ G$ can be thought of as the quotient of X^{ss} by the equivalence relation that $p \sim q$ if $\overline{G \cdot p} \cap \overline{G \cdot q}$ is non-empty in X^{ss}. The role of the stable points X^s is that G has closed orbits on X^s, so a "geometric quotient" X^s/G exists, and this sits inside the GIT quotient $X /\!\!/ G$.

5.3. The Hilbert-Mumford criterion

Let us suppose as in the previous sections, that a complex reductive group $G \subset GL(n+1, \mathbf{C})$ acts on a projective variety $X \subset \mathbf{CP}^n$, where the action is induced by the natural action on $\mathbf{CP}^n$. In this section we will discuss a criterion for determining whether a given point $p \in X$ is stable or semistable.

For any $p \in X$, let us write $\hat{p} \in \mathbf{C}^{n+1} \setminus \{0\}$ for a lift of p with respect to the projection map $\mathbf{C}^{n+1} \setminus \{0\} \to \mathbf{CP}^n$. We will write $G \cdot \hat{x}$ for the G orbit of $\hat{x}$ in $\mathbf{C}^{n+1}$.

Proposition 5.14. (a) *A point $p \in X$ is semistable if and only if $0 \notin \overline{G \cdot \hat{p}}$.*

(b) *A point $p \in X$ is stable if and only if the orbit $G \cdot \hat{p}$ is closed in $\mathbf{C}^{n+1}$ and the stabilizer of p in G is finite.*

Sketch of proof. (a) If $p \in X$ is semistable, then there is a homogeneous G-invariant polynomial f of positive degree which does not vanish at $\hat{p}$. The G-invariance implies that f is a non-zero constant on the orbit closure $\overline{G \cdot \hat{p}}$, so the origin cannot be in this closure.

Conversely if $0 \notin \overline{G \cdot \hat{p}}$, then one can show that there exists a G-invariant homogeneous polynomial f distinguishing the disjoint, closed, G-invariant sets 0 and $\overline{G \cdot \hat{p}}$. This polynomial f does not vanish at $\hat{p}$, so p is semistable.

(b) Suppose first that $p \in X$ is stable. If $G \cdot \hat{p}$ is not closed, then the closure contains another orbit $G \cdot \hat{q}$, for some $q \in \overline{G \cdot p}$. Then necessarily $q \in X$, and q is semistable. This contradicts the fact that the orbit of p in X^{ss} is closed.

Conversely suppose that $G \cdot p$ is not closed in X^{ss}, and let $q \in \overline{G \cdot p} \cap X^{ss}$ such that $q \notin G \cdot p$. Then there is a non-constant homogeneous G-invariant polynomial f which does not vanish at p and q, and we can assume that $f = 1$ on $G \cdot \hat{p}$ and $G \cdot \hat{q}$. From this one shows that the closure of $G \cdot \hat{p}$ contains $G \cdot \hat{q}$ and in particular $G \cdot \hat{p}$ is not closed. □

Since we will need it later, we define a third notion of stability at this point.

Definition 5.15. A point $p \in X$ is polystable if the orbit $G \cdot \hat{p}$ is closed. Note that stable $\Rightarrow$ polystable $\Rightarrow$ semistable but the converses are false in general. We will see later that the closure of every semistable orbit contains a unique polystable orbit. In other words the GIT quotient $X /\!/ G$ can be thought of as parametrizing the polystable orbits.

The Hilbert-Mumford criterion essentially says that in order to check whether an orbit $G\cdot\hat{p}$ is closed, it is enough to check this for all one-parameter subgroups $\mathbf{C}^* \subset G$. In practice this is very useful since the action of a one-parameter group can always be diagonalized, and this makes it possible to do some explicit calculations. In addition the Hilbert-Mumford criterion will motivate the definition of K-stability.

Definition 5.16. Suppose that $\lambda : \mathbf{C}^* \hookrightarrow G$ is a one-parameter subgroup. For any $p \in X$, define the weight $\mu(x, \lambda)$ as follows. First, let $q \in X$ be the limit

$$q = \lim_{t\to 0} \lambda(t) \cdot p$$

(we will see below that this limit exists). The point q is necessarily fixed by the one-parameter subgroup λ, so there exists an integer w such that $\lambda(t) \cdot \hat{q} = t^w \hat{q}$ for all t. We define $\mu(p, \lambda) = -w$.

A useful way to think of this is the following. Given a one-parameter subgroup of G acting on $\mathbf{C}^{n+1}$, we can write $\mathbf{C}^{n+1}$ as a sum of weight spaces

$$\mathbf{C}^{n+1} = \bigoplus_{i=1}^{k} V(w_i),$$

where each w_i is an integer, $\lambda(t) \cdot v = t^{w_i} v$ for $v \in V(w_i)$, and $k \leqslant n + 1$. We can arrange that $w_1 < w_2 < \cdots < w_k$. Given $p \in X$, we can write $\hat{p} = \hat{p}_1 + \cdots + \hat{p}_k$, where $\hat{p}_i \in V(w_i)$. If l is the smallest index for which $\hat{p}_l$ is non-zero, then the limit $q = \lim_{t\to 0} \lambda(t) \cdot p$ is obtained by letting $\hat{q} = \hat{p}_l$. Then $\mu(p, \lambda) = -w_l$.

Theorem 5.17 (Hilbert-Mumford criterion, [**85**, Theorem 2.1]).

(a) $p \in X$ *is semistable* $\Leftrightarrow \mu(p, \lambda) \geqslant 0$ *for all one-parameter subgroups* λ.

(b) $p \in X$ *is polystable* $\Leftrightarrow \mu(p, \lambda) > 0$ *for all one-parameter subgroups* λ *for which* $\lim_{t\to 0} \lambda(t) \cdot p \notin G \cdot p$.

(c) $p \in X$ *is stable* $\Leftrightarrow \mu(p, \lambda) > 0$ *for all one-parameter subgroups* λ.

Remarks on the proof. One direction of the result is fairly straightforward. For instance for part (a) suppose that λ is a one-parameter subgroup such that $\mu(p, \lambda) < 0$. Following the discussion before the theorem, we can write $\hat{p} = \hat{p}_1 + \cdots + \hat{p}_k$ in terms of the weight spaces of λ. Then

$$\lambda(t) \cdot \hat{p} = t^{w_1}\hat{p}_1 + \cdots + t^{w_k}\hat{p}_k,$$

and $\mu(p, \lambda) < 0$ means that the smallest weight w_i for which $\hat{p}_i \neq 0$ is positive. This means that λ acts on $\hat{p}$ with only positive weights, and so

$$\lim_{t\to 0} \lambda(t) \cdot \hat{p} = 0.$$

Therefore $0 \in \overline{G \cdot \hat{p}}$, and so p cannot be semistable.

The difficult part of the theorem is to show the converse. One method is to reduce the problem to the case of a torus action, for which the statement can be checked directly (see Exercise 5.21). □

Example 5.18. This example is the algebro-geometric counterpart to Example 5.6, where we looked at the action of $SU(2)$ on n-tuples of points on $\mathbf{CP}^1$. Let

$$V_n = \{\text{homogeneous degree } n \text{ polynomials in } x, y\} \cong \mathbf{C}^{n+1},$$

and let $X = \mathbf{P}(V_n)$. By identifying a polynomial with its zero set on $\mathbf{CP}^1$, we can think of X as the space of unordered n-tuples of points on $\mathbf{CP}^1$.

The group $SL(2, \mathbf{C})$ acts on V_n by

$$(M \cdot P)(x, y) = P(M^{-1}(x, y)),$$

where $M \in SL(2, \mathbf{C}), P \in V_n$, and $M^{-1}(x, y)$ is the standard action of $SL(2, \mathbf{C})$ on $\mathbf{C}^2$. In terms of n-tuples of points, this action corresponds to moving the points around on $\mathbf{CP}^1$, using the usual action of $SL(2, \mathbf{C})$ on $\mathbf{CP}^1$.

Let us determine the stable points for this action. Let λ be a one-parameter subgroup of $SL(2, \mathbf{C})$. We can choose a basis u, v for $\mathbf{C}^2$ such that in this basis λ is given by

$$\lambda(t) = \begin{pmatrix} t^w & 0 \\ 0 & t^{-w} \end{pmatrix},$$

for some integer $w > 0$. The induced action on a polynomial $P(u, v) = a_0 u^n + a_1 u^{n-1} v + \cdots + a_n v^n$ is

$$(\lambda(t) \cdot P)(u, v) = t^{-nw} a_0 u^n + t^{-(n-2)w} a_1 u^{n-1} v + \cdots + t^{nw} a_n v^n.$$

Writing $[P] \in \mathbf{P}(V_n)$ for the point in projective space corresponding to P, we have

$$\begin{aligned}[\lambda(t) \cdot P] &= [t^{-nw} a_0 u^n + \cdots + t^{nw} a_n v^n] \\ &= [a_k u^{n-k} v^k + t^{2w} a_{k+1} u^{n-k-1} v^{k+1} + \cdots + t^{(2n-2k)w} a_n v^n],\end{aligned}$$

where k is the smallest index for which $a_k \neq 0$. Therefore

$$\lim_{t\to 0}[\lambda(t)\cdot P] = [a_k u^{n-k} v^k].$$

Since

$$\lambda(t)\cdot(a_k u^{n-k} v^k) = t^{2k-n} a_k u^{n-k} v^k,$$

we have $\mu([P],\lambda) = n-2k$. By the Hilbert-Mumford criterion we need $k < n/2$ for $[P]$ to be stable. Since k was the smallest index for which $a_k \neq 0$, this means that P is not divisible by $v^{n/2}$, i.e. in terms of the n-tuple of zeros of P, the point $[1:0]$ has multiplicity less than $n/2$. Choosing different one-parameter subgroups amounts to looking at different points, so we obtain that an n-tuple of points is stable if and only if no point is repeated $n/2$ times.

In a similar way one can determine that an n-tuple is semistable if and only if no point is repeated more than $n/2$ times. Finally an n-tuple is polystable if either it is stable or it consists of just 2 points with multiplicity $n/2$. If n is odd, then all three notions of stability coincide.

Recall that in Example 5.6 we saw that if we look at the action of $SU(2)$ on such n-tuples on $\mathbf{CP}^1$, then zeros of the moment map are those n-tuples of points whose center of mass is the origin (thinking of $\mathbf{CP}^1$ as the unit sphere $S^2 \subset \mathbf{R}^3$). The Kempf-Ness theorem which we will discuss in the next section implies that an n-tuple is polystable if and only if its $SL(2,\mathbf{C})$-orbit contains a zero of the moment map. In other words we can move an n-tuple of points on $\mathbf{CP}^1$ into a balanced position (with center of mass the origin) by an element in $SL(2,\mathbf{C})$ if and only if no $n/2$ points coincide or the n-tuple consists of just two points with multiplicity $n/2$. One direction of this is clear: if too many points coincide, then we certainly cannot make the center of mass be the origin.

5.4. The Kempf-Ness theorem

Suppose now that $M \subset \mathbf{CP}^n$ is a projective submanifold with a complex group $G \subset GL(n+1,\mathbf{C})$ acting on M. Let $K = G \cap U(n+1)$, and assume that $K \subset G$ is a maximal compact subgroup. This means, on the level of Lie algebras, that

$$\mathfrak{g} = \mathfrak{k} \oplus \sqrt{-1}\mathfrak{k}.$$

Recall that

$$\begin{aligned} \mu_U : \mathbf{CP}^n &\to \mathfrak{u}(n+1)^* \\ [Z_0 : \cdots : Z_n] &\mapsto \frac{\sqrt{-1}Z_i\overline{Z}_j}{|Z|^2} \end{aligned}$$

is a moment map for the $U(n+1)$-action on $\mathbf{CP}^n$. The restriction of this to M, projected to $\mathfrak{k}^*$, is a moment map

$$\mu : M \to \mathfrak{k}^*$$

for the action of K on M, with respect to the symplectic form given by the restriction of the Fubini-Study metric to M. Recall also that for $p \in M$ we let $\hat{p} \in \mathbf{C}^{n+1} \setminus \{0\}$ be a lift and say that p is polystable for the action of G if the G-orbit $G \cdot \hat{p} \subset \mathbf{C}^{n+1}$ is closed.

Theorem 5.19 (Kempf-Ness, [**85**, Theorem 8.3])**.** *A point $p \in M$ is polystable for the action of G if and only if the orbit $G \cdot p$ contains a zero of the moment map μ. Moreover if p is polystable, then $G \cdot p \cap \mu^{-1}(0)$ is a single K-orbit.*

Sketch of proof. Let us introduce the function

$$\begin{aligned} \mathcal{M} : G/K &\to \mathbf{R} \\ [g] &\mapsto \log |g \cdot \hat{p}|^2, \end{aligned}$$

where $|\cdot|$ is the Euclidean norm on $\mathbf{C}^{n+1}$ and $[g]$ denotes the coset gK. Note that since $K \subset U(n+1)$, the K-action preserves the norm, so $\mathcal{M}$ is well-defined.

The space G/K can be endowed with a Riemannian metric, so that it is a non-positively curved symmetric space. The geodesics are given by one-parameter subgroups $[e^{t\sqrt{-1}\xi}g]$ for $\xi \in \mathfrak{k}$ and $g \in G$. The two main points are the following:

(i) $[g]$ is a critical point of $\mathcal{M}$ if and only if $\mu(g \cdot x) = 0$.

(ii) $\mathcal{M}$ is convex along geodesics in G/K.

The orbit $G \cdot \hat{p}$ is closed precisely when the norm $|g \cdot \hat{p}|$ goes to infinity as g goes to infinity, and this corresponds to the function $\mathcal{M}$ being proper. Because of the convexity, this happens exactly when $\mathcal{M}$ has a critical point.

To see (i), we need to compute the derivative of $\mathcal{M}$. Fix a $g \in G$ and write $g \cdot \hat{p} = Z$ and choose a skew-Hermitian matrix $A \in \mathfrak{k}$. We have

$$\begin{aligned} \left.\frac{d}{dt}\right|_{t=0} \mathcal{M}(e^{t\sqrt{-1}A}g) &= \left.\frac{d}{dt}\right|_{t=0} \log |e^{t\sqrt{-1}A}g \cdot \hat{p}|^2 \\ &= \frac{-\sqrt{-1}Z^T\overline{AZ} + \sqrt{-1}(AZ)^T\overline{Z}}{|Z|^2} \\ &= \frac{2\sqrt{-1}\overline{Z}^T AZ}{|Z|^2} \\ &= -2\langle \mu(g \cdot p), A \rangle, \end{aligned} \tag{5.2}$$

where we used that A is skew-Hermitian and we are using the pairing $\langle A, B\rangle = -\mathrm{Tr}(AB)$. It follows that $[g]$ is a critical point of $\mathcal{M}$ if and only if $\mu(g\cdot p) = 0$.

To see (ii) we need to compute the second derivative:

$$\begin{aligned}\frac{d^2}{dt^2}\bigg|_{t=0} \mathcal{M}(e^{t\sqrt{-1}A}g) &= -2\,\frac{d}{dt}\bigg|_{t=0} \langle \mu(e^{t\sqrt{-1}A}g\cdot p), A\rangle \\ &= 2g_{FS}(A_{g\cdot p}, A_{g\cdot p}) \geqslant 0,\end{aligned} \tag{5.3}$$

where g_{FS} is the Fubini-Study metric restricted to M and by $A_{g\cdot p}$ we mean the tangent vector at $g\cdot p$ induced by the infinitesimal action of A. □

Exercise 5.20. With the notation of Exercise 5.8, note that $GL(n, \mathbf{C})$ also acts on M_n by conjugation (it does not preserve the metric though).

(a) Identify the closed orbits of this action.

(b) By the Kempf-Ness theorem, every closed $GL(n, \mathbf{C})$-orbit contains a $U(n)$-orbit of zeroes of the moment map μ. What linear algebra statement does this correspond to?

Exercise 5.21. Suppose that the torus $T = (\mathbf{C}^*)^k$ acts on a vector space V with an induced action on $\mathbf{P}(V)$. Decompose V into weight spaces

$$V = \bigoplus_{\xi} V(\xi),$$

where $\xi \in \mathfrak{t}^*$. For each $p \in \mathbf{P}(V)$ we can define the weight polytope $\Delta(p) \subset \mathfrak{t}^*$ by

$$\Delta(p) = \mathrm{span}\{\xi \in \mathfrak{t}^* \,:\, \hat{p} \in V \text{ has non-zero component in } V(\xi)\},$$

where $\hat{p} \in V$ is any lift of p. We have the following.

(a) p is stable if and only if $\Delta(p)$ contains the origin in its interior.

(b) p is semistable if and only if $0 \in \Delta(p)$.

(c) p is polystable if and only if the origin is contained in the relative interior of $\Delta(p)$ (relative to the affine subspace spanned by $\Delta(p)$).

(d) Deduce the Hilbert-Mumford criterion for torus actions.

(e) Let $H \subset \mathfrak{t}^*$ denote the affine subspace spanned by $\Delta(p)$. Then p is relatively stable if and only if the projection of the origin onto H (using the metric used in taking the norm of the moment map) is in the relative interior of $\Delta(p)$.

(f) The weight polytope $\Delta(p)$ is the image of the moment map for the action of the compact torus $(S^1)^k$ restricted to the T-orbit of p. Use this to verify the Kempf-Ness theorem for torus actions.

In the next chapter we will see that the scalar curvature of a Kähler metric arises as the moment map for an infinite-dimensional Hamiltonian action. By comparing formula (5.2) to the variation (4.2) of the Mabuchi functional, we see that the function $\mathcal{M}$ is the finite-dimensional analog of the Mabuchi functional from Section 4.3. There is also an analog of the Futaki invariant, given as follows. For $p \in M$, let $G_p \subset G$ be the stabilizer of p, and let $\mathfrak{g}_p$ be its Lie algebra. The group G_p acts on the line spanned by $\hat{p}$, and we denote the infinitesimal action by the Lie algebra homomorphism

$$F : \mathfrak{g}_p \to \mathbf{C}.$$

We can compute F in terms of the moment map. Writing $Z = \hat{p}$ again, if $A \in \mathfrak{g}_p$, then $AZ = F(A)Z$ by definition. Then

$$\langle \mu(p), A \rangle = \frac{-\sqrt{-1}\overline{Z}^T AZ}{|Z|^2} = -\sqrt{-1}F(A),$$

so

$$F(A) = \sqrt{-1}\langle \mu(p), A \rangle. \tag{5.4}$$

This formula should be compared with definition (4.1) of the Futaki invariant in order to see the analogy.

Remark 5.22. If we have a Hamiltonian action $K \curvearrowright (M, \omega)$, then the choice of equivariant moment map $\mu : M \to \mathfrak{k}^*$ is only unique up to adding an element in the center of $\mathfrak{k}^*$. The symplectic quotient in turn depends on the choice of moment map. On the algebraic side, if K acts by biholomorphisms and $L \to M$ is a Hermitian line bundle with curvature form $2\pi\omega$, then the ambiguity in the choice of a moment map corresponds to a choice in how we lift the action of K (and its complexification G) to the space of sections of L^k. In Section 5.2 this choice is fixed by thinking of our group as a subgroup of $GL(n+1, \mathbf{C})$. More generally it would be enough to have a representation $G \to GL(n+1, \mathbf{C})$, lifting the action of G on M, which corresponds to a map $G \to PGL(n+1, \mathbf{C})$. A lift of the G action to the sections of L^k is called a linearization of the action.

In the Kempf-Ness theorem it is crucial that the choice of moment map is consistent with the choice of linearization. More invariantly this consistency can be expressed as follows. Suppose that $\xi \in \mathfrak{k}$ induces the holomorphic Hamiltonian vector field v on M with Hamiltonian function given by $H = \langle \mu, \xi \rangle$. Then the induced (dual) action on $H^0(M, L)$ is given by

$$\xi \cdot s = \nabla_{-v} s - 2\pi\sqrt{-1}Hs. \tag{5.5}$$

A calculation shows that this is consistent with our choice of moment map for $U(n+1)$ acting on $\mathbf{CP}^n$.

5.5. Relative stability

Suppose that we are in the setting of the Kempf-Ness theorem, but in the more invariant formulation of Remark 5.22. Thus we have a Kähler manifold (M, ω) together with an ample line bundle $L \to M$ with $c_1(L) = [\omega]$. We assume that we have a Hamiltonian action of a compact Lie group K on (M, ω) with moment map $\mu : M \to \mathfrak{k}^*$ and that the complexified action of $G = K^c$ has a lift to the total space of L.

The Kempf-Ness theorem characterizes orbits of the complexified group G that contain zeros of the moment map in terms of stability. In the study of extremal metrics it will be necessary to extend this to complex orbits which contain critical points of the norm squared $\|\mu\|^2$ of the moment map. In this section we will extend the Kempf-Ness theorem to characterize these orbits, following [**67**] and [**107**].

In order to define the norm squared of μ, we need to choose an inner product $\langle\cdot,\cdot\rangle$ on $\mathfrak{k}$ which is invariant under the adjoint action. Using this inner product we obtain an identification $\mathfrak{k} \cong \mathfrak{k}^*$ and so we can think of the moment map as $\mu : M \to \mathfrak{k}$.

Lemma 5.23. *A point p is a critical point of $\|\mu\|^2$ if and only if the element $\mu(p) \in \mathfrak{k}$ is in the stabilizer of p.*

Proof. We have

$$d\langle\mu,\mu\rangle(p) = 2\langle d\mu(p), \mu(p)\rangle = -2\omega_p(\rho(\mu)(p), \cdot),$$

where $\rho : \mathfrak{k} \to \mathrm{Vect}(M)$ is the infinitesimal action. Since the symplectic form is non-degenerate, we find that p is a critical point if and only if $\rho(\mu)(p) = 0$. □

We will also need the following, which is analogous to Lichnerowicz's result, Proposition 4.18, on the automorphism group of a cscK manifold.

Lemma 5.24. *Suppose that $\mu(p) = 0$. Then the stabilizer $\mathfrak{g}_p$ is the complexification of the stabilizer $\mathfrak{k}_p$.*

Proof. Let us write $\rho : \mathfrak{k} \oplus i\mathfrak{k} = \mathfrak{g} \to T_pM$ for the infinitesimal action and J for the complex structure on M. If $\xi, \eta \in \mathfrak{k}$, then we have

$$\begin{aligned}
|\rho(\xi + i\eta)|^2 &= \omega_p(\rho(\xi + i\eta), J\rho(\xi + i\eta)) \\
&= \omega_p(\rho(\xi) + J\rho(\eta), J\rho(\xi) - \rho(\eta)) \\
&= |\rho(\xi)|^2 + |\rho(\eta)|^2 - 2\omega_p(\rho(\xi), \rho(\eta)) \\
&= |\rho(\xi)|^2 + |\rho(\eta)|^2 + 2\langle d\mu_p(\rho(\eta)), \xi\rangle \\
&= |\rho(\xi)|^2 + |\rho(\eta)|^2 + 2\langle[\eta, \mu(p)], \xi\rangle,
\end{aligned}$$

where we used the equivariance of the moment map in the last step. Since we are assuming that $\mu(p) = 0$, it follows that

$$|\rho(\xi + i\eta)|^2 = |\rho(\xi)|^2 + |\rho(\eta)|^2.$$

In particular whenever $\xi + i\eta \in \mathfrak{g}_p$, we have $\xi, \eta \in \mathfrak{k}_p$, which is what we wanted to prove. □

For the action of a subgroup $H \subset K$ a natural moment map

$$\mu_H : M \to \mathfrak{h}$$

is given by composing $\mu : M \to \mathfrak{k}$ with the orthogonal projection $\mathfrak{k} \to \mathfrak{h}$. The idea in the definition of relative stability is that if $\mathfrak{h}$ is orthogonal to the stabilizer of p and p is a critical point of $\|\mu\|^2$, then it follows that $\mu_H(p) = 0$ and we can apply the Kempf-Ness theorem. It turns out that instead of the whole stabilizer of p it is better to work with a maximal torus $T \subset K_p$. Let $\mathfrak{t} \subset \mathfrak{k}_p$ be its Lie algebra. We then define the subalgebra

$$\mathfrak{k}_{T^\perp} = \{\xi \in \mathfrak{k} \,:\, [\xi, \eta] = 0,\ \langle \xi, \eta \rangle = 0 \text{ for all } \eta \in \mathfrak{t}\}$$

and let $\mathfrak{g}_{T^\perp}$ be its complexification. One can show that these subalgebras correspond to closed subgroups of K, G (see Lemma 1.3.2 in [**106**]) which we will denote by $K_{T^\perp}, G_{T^\perp}$. With these definitions we have the following.

Theorem 5.25. *Suppose that the complexification T^c is a maximal torus in the stabilizer G_p. Then the G-orbit of p contains a critical point of $\|\mu\|^2$ if and only if p is GIT stable for the action of $G_{T^\perp}$.*

Proof. Assume first that p is stable for the action of $G_{T^\perp}$. The Kempf-Ness theorem implies that there is a point $q \in G_{T^\perp} \cdot p$ such that

$$\mu_{T^\perp}(q) = 0,$$

where $\mu_{T^\perp}$ is the projection of μ onto $\mathfrak{k}_{T^\perp}$. Since elements in $G_{T^\perp}$ commute with T and $T \subset K_p$, we know that T fixes q as well. Since the moment map is K-equivariant, it follows that $\mu(q)$ commutes with $\mathfrak{t}$. Since the projection of $\mu(q)$ orthogonal to $\mathfrak{t}$ vanishes, it then follows that $\mu(q) \in \mathfrak{t}$, and in particular $\mu(q)$ is in the stabilizer of q. By Lemma 5.23 it follows that q is a critical point of $\|\mu\|^2$.

Now suppose that $q = g^{-1}p$ is a critical point of $\|\mu\|^2$ for some $g \in G$. We define a new symplectic form $\widetilde{\omega} = (g^{-1})^*\omega$, which is invariant under the action of the compact group $\widetilde{K} = gKg^{-1}$. A moment map for the action of $\widetilde{K}$ on $(M, \widetilde{\omega})$ is given by

$$\widetilde{\mu}(x) = \mathrm{Ad}_g \mu(g^{-1}x) \in \mathrm{Ad}_g \mathfrak{k},$$

where we identify the Lie algebra $\widetilde{\mathfrak{k}}$ of $\widetilde{K}$ with $\mathrm{Ad}_g\mathfrak{k}$. In particular, by the invariance of the inner product, we have

$$\|\widetilde{\mu}(x)\|^2 = \|\mu(g^{-1}x)\|^2.$$

It follows that p is a critical point of $\|\widetilde{\mu}\|^2$, and so $\widetilde{\mu}(p) \in \widetilde{\mathfrak{k}}_p$ by Lemma 5.23. The equivariance of the moment map implies that $\widetilde{\mu}(p)$ is in the center of the stabilizer of p, so if $\widetilde{T} \subset \widetilde{K}_p$ is any maximal torus, then we have $\widetilde{\mu}(p) \in \widetilde{\mathfrak{t}}$. It follows that the projection of $\widetilde{\mu}(p)$ onto $\widetilde{\mathfrak{k}}_{\widetilde{T}^\perp}$ vanishes, so by the Kempf-Ness theorem p is stable for the action of $G_{\widetilde{T}^\perp}$. The remaining problem is to replace $\widetilde{T}$ with the original maximal torus $T \subset K_p$.

For this last step we first note that $\widetilde{T}^c$ is a maximal torus in G_p. Indeed we have seen that p is stable for the action of $G_{\widetilde{T}^\perp}$, so Lemma 5.24 implies that the stabilizer of p in $G_{\widetilde{T}^\perp}$ is the complexification of its stabilizer in $\widetilde{K}_{\widetilde{T}^\perp}$. The latter is trivial, using that $\widetilde{T} \subset \widetilde{K}_p$ is a maximal torus, so it follows that p has trivial stabilizer in $G_{\widetilde{T}^\perp}$. The Lie algebra of $G_{\widetilde{T}^\perp}$ is

$$\mathfrak{g}_{\widetilde{T}^\perp} = \{\xi \in \mathfrak{g} : [\xi, \eta] = 0, \quad \langle \xi, \eta \rangle = 0 \text{ for all } \eta \in \widetilde{\mathfrak{t}}\}, \tag{5.6}$$

and so it follows that the elements in the stabilizer $\mathfrak{g}_p$ that commute with $\widetilde{\mathfrak{t}}$ are all contained in $\widetilde{\mathfrak{t}}^c$. This means that $\widetilde{T}^c$ is a maximal torus in G_p.

By our assumption T^c is also a maximal torus in G_p, so there is an element $h \in G_p$ such that $T^c = h\widetilde{T}^c h^{-1}$. It follows that

$$G_{T^\perp} = hG_{\widetilde{T}^\perp}.$$

It is then clear from Proposition 5.14 that the stability of p for the action of $G_{\widetilde{T}^\perp}$ implies the stability for the action of $G_{T^\perp}$. □

From the last part of the proof it is clear that the compact group K does not play a role in stability for the action of $G_{T^\perp}$, but rather we simply need a maximal torus $T \subset G_p$, and we look at stability for the action of the group $G_{T^\perp}$ with Lie algebra defined just as in equation (5.6). The difference is that here T is a complex group. This leads to the following definition.

Definition 5.26. A point p is *relatively stable* for the action of G if p is stable for the action of $G_{T^\perp}$, where $T \subset G_p$ is any maximal torus. By Theorem 5.25 this is equivalent to saying that the orbit $G \cdot p$ contains a critical point of $\|\mu\|^2$.

Example 5.27. Let us examine the diagonal action of $SU(2) \curvearrowright \mathrm{Sym}^n\mathbf{CP}^1$ from Example 5.6 again. As before, a moment map for the action is given by

$$\mu(x_1, \ldots, x_n) = x_1 + \cdots + x_n,$$

where $\mathbf{CP}^1$ is identified with the unit sphere in $\mathbf{R}^3$ as a coadjoint orbit. We furthermore identify $\mathbf{R}^3$ with its dual using the Euclidean inner product. In terms of this identification a vector $\xi \in \mathbf{R}^3 = \mathfrak{su}(2)$ corresponds to rotation of $\mathbf{CP}^1$ about the axis spanned by ξ. It follows from Lemma 5.23 that critical points of $\|\mu\|^2$ in this case are those n-tuples $(x_1, \ldots, x_n)$ which are concentrated at two antipodal points, in addition to the zeros of the moment map. The relatively stable n-tuples, which are not stable, are then simply those n-tuples that are concentrated at two points.

Exercise 5.28. Show that for an S^1-action on $\mathbf{CP}^n$ generated by a Hermitian matrix A, the relatively stable points that are not zeros of the moment map correspond to the eigenvectors of A with non-zero eigenvalue.

Chapter 6

K-stability

In this chapter we will first describe how the scalar curvature of a Kähler metric arises as a moment map for an infinite-dimensional Hamiltonian action. This not only sheds light on the developments in Chapter 4 but is also the motivation for the notion of K-stability, as an analogous condition to the Hilbert-Mumford criterion in geometric invariant theory. The Calabi functional becomes the norm squared of the moment map, and we will introduce relative K-stability in analogy with the results in Section 5.5. A computation with the ruled surface studied in Section 4.4 will illustrate the relation between relative K-stability and the existence of extremal metrics. We will conclude with a discussion of K-stability for toric manifolds, giving the algebro-geometric side of the development in Section 4.5.

6.1. The scalar curvature as a moment map

In this section we will see that the scalar curvature can be viewed as a moment map, as was discovered by Fujiki [**53**] and Donaldson [**42**]. We will only sketch the construction, but the details can be found in [**42**] and also in Tian [**113**].

Let (M, ω) be a symplectic manifold. This means that ω is a closed, non-degenerate 2-form. For simplicity we will assume that $H^1(M, \mathbf{R}) = 0$. Recall that an almost complex structure on M is an endomorphism $J : TM \to TM$ such that $J^2 = -\mathrm{Id}$, where Id is the identity map. We say that an almost complex structure J is compatible with ω if the tensor

$$g_J(X, Y) = \omega(X, JY)$$

is symmetric and positive definite, i.e. it defines a Riemannian metric. If J is integrable, then (M, J) is a complex manifold and the metric g_J is Kähler.

Define the infinite-dimensional space

$$\mathscr{J} = \{\text{almost complex structures on } M \text{ compatible with } \omega\}.$$

The tangent space at a point J is given by

$$T_J\mathscr{J} = \left\{ \begin{array}{l} A : TM \to TM \text{ such that } AJ + JA = 0 \\ \text{and } \omega(X, AY) = \omega(Y, AX) \text{ for all } X, Y \end{array} \right\}.$$

If $A \in T_J\mathscr{J}$, then also $JA \in T_J\mathscr{J}$, and this defines a complex structure on $\mathscr{J}$. We can also define an inner product

$$\langle A, B\rangle_J = \int_M \langle A, B\rangle_{g_J} \frac{\omega^n}{n!},$$

for $A, B \in T_J\mathscr{J}$, which gives rise to a Hermitian metric on $\mathscr{J}$. Combining these structures we have a Kähler form on $\mathscr{J}$, given at the point J by

$$\Omega_J(A, B) = \langle JA, B\rangle_J.$$

We now let

$$\mathscr{G} = \{\text{the group of Hamiltonian symplectomorphisms of } (M, \omega)\}.$$

These are the time-one flows of time dependent Hamiltonian vector fields. Using the Hamiltonian construction, the Lie algebra $\mathrm{Lie}(\mathscr{G})$ can be identified with the functions on M with zero integral, $C_0^\infty(M)$. The group $\mathscr{G}$ acts on $\mathscr{J}$ by pulling back complex structures, and this action preserves the Kähler form Ω.

Theorem 6.1 (Fujiki [**53**] and Donaldson [**42**]). *The action of $\mathscr{G}$ on $\mathscr{J}$ is Hamiltonian, and a moment map is given by*

$$\begin{aligned} \mu : \mathscr{J} &\to \mathrm{Lie}(\mathscr{G})^* \\ J &\mapsto S(J) - \hat{S}, \end{aligned}$$

where $S(J)$ is the scalar curvature of g_J when J is integrable and $\hat{S}$ is the average of $S(J)$, which does not depend on J. The function $S(J) - \hat{S}$ is thought of as an element of the dual of $\mathrm{Lie}(\mathscr{G}) \cong C_0^\infty(M)$ through the L^2-product on M.

Note that if J is not integrable, then $S(J)$ is the "Hermitian scalar curvature" of g_J, which is not the same as the Riemannian scalar curvature. In any case, the theorem means that finding constant scalar curvature Kähler metrics amounts to finding integrable J with $\mu(J) = 0$.

Let us unwind what it means for μ to be a moment map. For any $J \in \mathscr{J}$ there are two linear operators

$$\begin{aligned} P &: C_0^\infty(M) \to T_J\mathscr{J}, \\ Q &: T_J\mathscr{J} \to C_0^\infty(M). \end{aligned}$$

The map P is given by the infinitesimal action of Lie($\mathscr{G}$) on $\mathcal{J}$. This can be written as

$$P(H) = L_{X_H}J,$$

where L_{X_H} is the Lie derivative with respect to the Hamiltonian vector field X_H. The map Q is the derivative of the map taking J to the Hermitian scalar curvature of g_J, so we can write

$$Q(A) = DS_J(A).$$

To say that μ is a moment map simply means that for all $A \in T_J\mathscr{J}$ and $H \in C_0^\infty(M)$ we have

$$\langle Q(A), H\rangle_J = -\Omega_J(P(H), A) = \langle JA, P(H)\rangle_{L^2},$$

where on the right we just have the L^2-product on functions. In other words the theorem means that $Q^* = -JP$ for the adjoint Q^* of Q.

We can also see how extremal metrics arise in this picture. Namely, using the L^2-product on the Lie algebra $C_0^\infty(M)$, we have

$$\|\mu(J)\|^2 = \int_M (S(J) - \hat{S})^2\,\omega^n, \tag{6.1}$$

so the norm squared of the moment map agrees up to a constant with the Calabi functional from Chapter 4. In particular, extremal metrics are critical points of $\|\mu\|^2$. Lemma 5.23 in turn tells us that J is a critical point of $\|\mu\|^2$ if and only if $\mu(J)$, thought of as an element of Lie($\mathscr{G}$), fixes the point J. In the notation above, this means that at the point $J \in \mathscr{J}$ we have $P(S(J) - \hat{S}) = 0$, i.e.

$$L_{X_{S(J)}}J = 0.$$

This equation expresses that the Hamiltonian vector field $X_{S(J)}$, or equivalently the gradient $\nabla S(J)$, preserves the complex structure J, which is the same as the Euler-Lagrange equation for extremal metrics that we obtained in Theorem 4.2.

Note that in earlier chapters we were always fixing a complex structure since we were working on a fixed complex manifold and we were varying the Kähler metric ω. Now we seem to be doing something rather different, fixing the form ω and varying the complex structure instead. These two points of view can be related to each other as follows. For any symplectic form ω and compatible complex structure J let us write $g(J,\omega)$ for the corresponding Kähler metric. If J, J' are two complex structures and $J' = f^*J$ for a diffeomorphism J, then

$$g(J',\omega) = f^*g(J, (f^{-1})^*\omega). \tag{6.2}$$

If $f \in \mathscr{G}$, then this means that the metrics $g(J,\omega)$ and $g(J',\omega)$ are isometric. To obtain something non-trivial, we need to consider the complexification

$\mathscr{G}^c$ of $\mathscr{G}$. While this complexified group does not exist, we can at least complexify the Lie algebra

$$\mathrm{Lie}(\mathscr{G}^c) = C_0^\infty(M, \mathbf{C}),$$

and since $\mathscr{J}$ has a complex structure, we can also naturally complexify the infinitesimal action. This complexified infinitesimal action gives rise to a foliation on $\mathscr{J}$ whose leaves can be thought of as the orbits of $\mathscr{G}^c$.

Claim 6.2. *If $J \in \mathscr{J}$ is integrable, then the $\mathscr{G}^c$-orbit of J (or rather just the orbit of the imaginary part of $\mathscr{G}^c$) can be identified with the set of Kähler metrics on (M, J) in the class $[\omega]$.*

To see this at an infinitesimal level, let $J \in \mathscr{J}$ be integrable, suppose that $H \in C_0^\infty(M)$, and let us see what the action of $\sqrt{-1}H$ looks like on $T_J\mathscr{J}$. When J is integrable, then

$$JP(H) = JL_{X_H}J = L_{JX_H}J,$$

so infinitesimally the action of $\sqrt{-1}H$ means flowing along the vector field JX_H. By observation (6.2) we obtain the same metric this way as if we fix J instead and flow ω along the vector field $-JX_H$. The variation of ω is then

$$-L_{JX_H}\omega = -d\iota_{JX_H}\omega = 2\sqrt{-1}\partial\bar{\partial}H, \tag{6.3}$$

and so flowing along this vector field amounts to moving ω in its Kähler class.

The upshot of all this is that, at least formally, the problem of finding a cscK metric in the Kähler class $[\omega]$ on the complex manifold (M, J) is equivalent to finding a zero of the moment map μ for the action of $\mathscr{G}$, in the orbit J under the complexified action. In the finite-dimensional case the Kempf-Ness theorem characterizes the orbits with zeros of the moment map in terms of GIT stability, which can be tested using one-parameter subgroups according to the Hilbert-Mumford criterion. One-parameter subgroups heuristically correspond to geodesic rays in the space of Kähler metrics. We know little about these in general, but they can be used to define a notion of stability; see Chen [**30**]. Instead of this, one way to think of K-stability is to approximate the space of Kähler metrics in the class $[\omega]$ by algebraic metrics obtained by pulling back the Fubini-Study metric under embeddings $M \subset \mathbf{P}^N$. This is an idea which goes back to Yau [**123**]. One can then use algebraic geometry to study degenerations of M inside projective space under one-parameter subgroups of $GL(N+1)$. In the limit as $N \to \infty$, we recover the infinite-dimensional picture, at least heuristically. One way in which this approximation can be made rigorous is through the Bergman kernel, which we will study in Chapter 7.

6.2. The Hilbert polynomial and flat limits

Recall that for a projective variety $X \subset \mathbf{CP}^n$ we defined the homogeneous coordinate ring

$$R(X) = \mathbf{C}[x_0, \dots, x_n]/I,$$

where I is the ideal generated by the homogeneous polynomials which vanish on X. This is a graded ring whose degree d piece $R_d(X)$ is the image of the degree d polynomials under the quotient map. Each $R_d(X)$ is a finite-dimensional vector space, and the Hilbert function of X is defined by

$$H_X(d) = \dim R_d(X).$$

A fundamental result is that there is a polynomial $P_X(d)$, called the Hilbert polynomial, such that $H_X(d) = P_X(d)$ for sufficiently large d. The degree of P_X is the dimension of X. The polynomial P_X should be thought of as an invariant of X, and one of its crucial properties is that it does not change if we vary X in a nice enough family (the technical condition is that the family is "flat").

The Hilbert function can be defined more generally for any homogeneous ideal $I \subset \mathbf{C}[x_0, \dots, x_n]$ by letting

$$H_I(d) = \dim \left(\mathbf{C}[x_0, \dots, x_n]/I\right)_d,$$

where again we are taking the image of the degree d polynomials under the quotient map. Once again one can show that for large enough d, the Hilbert function $H_I(d)$ coincides with a polynomial $P_I(d)$.

We will now give a very special example of a flat family, which will be enough for our needs. Suppose that $I \subset \mathbf{C}[x_0, \dots, x_n]$ is a homogeneous ideal and that we have a one-parameter subgroup $\lambda : \mathbf{C}^* \hookrightarrow GL(n+1, \mathbf{C})$. For any polynomial f, we can define $\lambda(t) \cdot f$ by

$$(\lambda(t) \cdot f)(x_0, \dots, x_n) = f\big(\lambda(t^{-1}) \cdot (x_0, \dots, x_n)\big),$$

and it is easy to check that

$$I_t = \{\lambda(t) \cdot f \mid f \in I\}$$

is also a homogeneous ideal in $\mathbf{C}[x_0, \dots, x_n]$. Geometrically the vanishing set of I_t is obtained by applying $\lambda(t)$ to the vanishing set of I.

Definition 6.3. The flat limit $I_0 = \lim_{t\to 0} I_t$ is defined as follows. We can decompose any $f \in I$ as $f = f_1 + \cdots + f_k$ into elements in distinct weight spaces for the $\mathbf{C}^*$-action λ on $\mathbf{C}[x_0, \dots, x_n]$. Let us write $\mathrm{in}(f)$ for the element f_i with the smallest weight, which we can think of as the "initial term" of f. Then I_0 is the ideal generated by the set of initial terms $\{\mathrm{in}(f) \mid f \in I\}$.

For any ideal I let us write $(I)_d$ for the degree d piece of I. Then one can check that for our flat limit, the degree d piece $(I_0)_d$ of I_0 is the vector space spanned by $\{\mathrm{in}(f) \mid f \in (I)_d\}$. From this it is not hard to see that $\dim(I_0)_d = \dim(I)_d$ for each d, so the Hilbert polynomials of I and I_0 are the same. In fact in this case even the Hilbert function is preserved in the limit, but that is not true for more general flat limits.

Example 6.4. A simple example is letting $I \subset \mathbf{C}[x, y, z]$ be the ideal $I = (xz - y^2)$, i.e. the ideal generated by the polynomial $xz - y^2$. The corresponding projective variety is a conic in $\mathbf{CP}^2$. Let us take the $\mathbf{C}^*$-action given by $\lambda(t) \cdot (x, y, z) = (tx, t^{-1}y, z)$. The dual action on functions gives

$$\lambda(t) \cdot (xz - y^2) = t^{-1}xz - t^2y^2.$$

The initial term is $\mathrm{in}(xz - y^2) = xz$, and so the flat limit is

$$\lim_{t\to 0} \lambda(t) \cdot I = (xz).$$

The variety corresponding to (xz) is two lines intersecting in a point. In other words, when taking the limit, the conic breaks up into two intersecting lines. While this limit is not irreducible, it is still the union of two projective varieties.

Note that in general by just taking the initial terms of a set of generators of the ideal, we might get a smaller ideal than the flat limit. In the example we are looking at here, we can check that the Hilbert polynomial of (xz) equals the Hilbert polynomial of $(xz - y^2)$, so (xz) has to be the flat limit. More generally one can use Gröbner bases to do these calculations.

Example 6.5. For a similar example let us take $I = (xz - y^2)$ again, but let $\lambda(t) \cdot (x, y, z) = (t^{-1}x, ty, z)$. Then

$$\lambda(t) \cdot (xz - y^2) = txz - t^{-2}y^2,$$

so now the initial term is $-y^2$, and the flat limit is

$$\lim_{t\to 0} \lambda(t) \cdot I = (y^2).$$

The zero set of (y^2) is a line in $\mathbf{CP}^2$, but it should be thought of as having multiplicity 2, or as being "thickened". The quotient ring $\mathbf{C}[x, y, z]/(y^2)$ has nilpotents, and the corresponding geometric object is a projective scheme.

The flat limits that we are considering arise when we try to form a GIT quotient of the space of all projective subvarieties in $\mathbf{CP}^n$. More precisely one can show that given a polynomial P, there is a projective scheme $\mathrm{Hilb}_{P,n}$, called the Hilbert scheme, parametrizing all projective subschemes of $\mathbf{CP}^n$ with Hilbert polynomial P. The idea is that if a homogeneous ideal $I \subset \mathbf{C}[x_0, \ldots, x_n]$ has Hilbert polynomial P, then there is a large number d

(depending on P, n) such that the scheme corresponding to I is determined by the degree d piece of I, which we still denote by $(I)_d$ (technically for this one needs to restrict attention to "saturated" ideals—saturating an ideal does not change the corresponding scheme). This $(I)_d$ is simply a linear subspace of the degree d polynomials $\mathbf{C}[x_0, \ldots, x_n]_d$ such that

$$\dim \mathbf{C}[x_0, \ldots, x_n]_d/(I)_d = P(d).$$

Since the degree d piece of I determines the scheme, we obtain a map from the set of schemes with Hilbert polynomial P to a certain Grassmannian of subspaces of a finite-dimensional vector space. Roughly speaking, the image of this map is the Hilbert scheme (although it has more structure than just being a subset).

The "moduli space" of projective varieties (or schemes) in $\mathbf{CP}^n$ with a given Hilbert polynomial should then be the GIT quotient

$$\mathrm{Hilb}_{P,n} /\!\!/ SL(n+1, \mathbf{C})$$

since acting by $SL(n+1, \mathbf{C})$ simply changes the embedding of a variety, not the variety itself. If we try to use the Hilbert-Mumford criterion to determine whether a given variety is stable (or semistable), then we naturally arrive at the notion of a flat limit under a $\mathbf{C}^*$-action which we defined above.

6.3. Test-configurations and K-stability

In this section we will introduce the notion of K-stability. This was originally defined by Tian [**112**] and conjectured to characterize the existence of a Kähler-Einstein metric on a manifold with positive first Chern class. A more refined, algebro geometric definition was introduced by Donaldson [**44**], which he conjectured to characterize the existence of a cscK metric. It is this definition which we will use.

The definition of K-stability is inspired by the Hilbert-Mumford criterion for stability in GIT, which we discussed in Section 5.3. Throughout this section we will work with a projective manifold $M \subset \mathbf{CP}^N$ with $\dim M = n$, although it is more natural to think of the pair (M, L), where $L = \mathcal{O}(1)|_M$. In general such a pair (M, L) of a projective manifold together with an ample line bundle L is called a polarized manifold.

Definition 6.6. A test-configuration for (M, L) of exponent $r > 0$ consists of an embedding $M \hookrightarrow \mathbf{CP}^{N_r}$ using a basis of sections of L^r and a $\mathbf{C}^*$-subgroup of $GL(N_r + 1, \mathbf{C})$.

Given a test-configuration with $\lambda : \mathbf{C}^* \hookrightarrow GL(N_r + 1, \mathbf{C})$ being the $\mathbf{C}^*$-subgroup, we obtain a family of submanifolds $M_t \subset \mathbf{CP}^{N_r}$, with

$M_t = \lambda(t) \cdot M$. This family can be extended across $t = 0$ by taking the flat limit

$$M_0 = \lim_{t \to 0} M_t$$

according to Definition 6.3. Usually the definition of a test-configuration is formulated in terms of the resulting flat $\mathbf{C}^*$-equivariant family over $\mathbf{C}$. By construction the flat limit M_0 is preserved by the $\mathbf{C}^*$-action λ, and by analogy with the Hilbert-Mumford criterion, we need to define a weight for this action. This weight is given by the Donaldson-Futaki invariant.

To define the weight, suppose that $X \subset \mathbf{CP}^N$ is a subscheme, invariant under a $\mathbf{C}^*$-action λ. Algebraically this means that we have a homogeneous ideal

$$I \subset \mathbf{C}[x_0, \ldots, x_N],$$

which is preserved by the dual action of λ. It follows that there is an induced $\mathbf{C}^*$-action on the homogeneous coordinate ring

$$R = \mathbf{C}[x_0, \ldots, x_N]/I$$

and each degree k piece R_k is invariant. Let us write $d_k = \dim R_k$ for the Hilbert function of X, and let w_k be the total weight of the action on R_k. As we discussed in Section 6.2, for large k, d_k equals a polynomial, so we have

$$d_k = a_0 k^n + a_1 k^{n-1} + O(k^{n-2}),$$

for some constants a_0, a_1. The degree n is the dimensions of X. Similarly w_k equals a polynomial of degree $n+1$ for large k, so we can define constants b_0 and b_1 by

$$w_k = b_0 k^{n+1} + b_1 k^n + O(k^{n-1}).$$

Definition 6.7. The Donaldson-Futaki invariant of the $\mathbf{C}^*$-action λ on X is defined to be

$$F(X, \lambda) = \frac{a_1}{a_0} b_0 - b_1.$$

Note that in the literature sometimes the formula has the opposite signs because of varying conventions on whether one takes the dual action or not. We will see in Section 7.4 that if X is smooth and the $\mathbf{C}^*$-action is induced by a holomorphic vector field, then the Donaldson-Futaki invariant coincides with the differential geometric Futaki invariant introduced in Section 4.2.

Before defining K-stability we need to define when we consider a test-configuration to be trivial. Originally a test-configuration was defined to be trivial if the central fiber M_0 is biholomorphic to M. It was pointed out by Li-Xu [**75**] that this is not enough in general unless one restricts attention to those test-configurations whose total space is "normal". This is a condition on the type of singularities that can occur. Instead we give an alternative definition, relying on the norm of a test-configuration. Using the same notation as above, suppose that we have a $\mathbf{C}^*$-action λ on a subscheme $X \subset \mathbf{CP}^N$, and let us write A_k for the infinitesimal generator of the $\mathbf{C}^*$-action on the degree k piece R_k of the homogeneous coordinate ring. Then $\mathrm{Tr}(A_k) = w_k$ in the notation above. Similarly to d_k and w_k, the function $\mathrm{Tr}(A_k^2)$ is a polynomial of degree $n+2$, and we define c_0 by

$$\mathrm{Tr}(A_k^2) = c_0 k^{n+2} + O(k^{n+1}).$$

The norm $\|\lambda\|$ of the $\mathbf{C}^*$-action λ is defined to be

$$\|\lambda\|^2 = c_0 - \frac{b_0^2}{a_0},$$

where a_0, b_0 are as above. In other words, $\|\lambda\|^2$ is the leading term in

$$\mathrm{Tr}\left(A_k - \frac{\mathrm{Tr}(A_k)}{d_k}\mathrm{Id}\right)^2 = \|\lambda\|^2 k^{n+2} + O(k^{n+1}). \tag{6.4}$$

We can now give the definition of K-stability.

Definition 6.8. Let (M, L) be a polarized manifold. Given a test-configuration χ for (M, L), let us also write χ for the induced $\mathbf{C}^*$-action on the central fiber, so we have the norm $\|\chi\|$ and the Donaldson-Futaki invariant $F(\chi) = F(M_0, \chi)$.

The pair (M, L) is K-semistable if for every test-configuration χ we have $F(\chi) \geqslant 0$. If in addition $F(\chi) > 0$ whenever $\|\chi\| > 0$, then (M, L) is K-stable.

One version of the central conjecture in the field is the following.

Conjecture 6.9 (Yau-Tian-Donaldson)**.** *Let (M, L) be a polarized manifold, and suppose that M has discrete holomorphic automorphism group. Then M admits a cscK metric in $c_1(L)$ if and only if (M, L) is K-stable.*

In Section 6.4 we will discuss a version of this conjecture applicable when M has holomorphic vector fields and where cscK metrics are replaced by extremal metrics. In Remark 6.18 we will briefly discuss an example due to Apostolov–Calderbank–Gauduchon–Tønnesen-Friedman [**2**], which suggests that the notion of K-stability needs to be strengthened, so as written the conjecture should be thought of more as a guiding principle. In Section 6.6 we will discuss one possible way of strengthening K-stability.

Exercise 6.10. Suppose that (M, L_M) and (N, L_N) are polarized manifolds, and let $L \to M \times N$ be given by $L = \pi_1^* L_M \otimes \pi_2^* L_N$, where π_1, π_2 are the projections onto the two factors. Assuming that $(M \times N, L)$ is K-stable, show that (M, L_M) and (N, L_N) are K-stable and that the validity of Conjecture 6.9 implies the converse.

One of our goals in this book is to explain the proof of one direction of the conjecture, due to Stoppa [**101**], which built on the work of Donaldson [**46**] and Arezzo-Pacard [**3**] (see also Mabuchi [**79**]). In the Kähler-Einstein case the result is due to Tian [**112**] and Paul-Tian [**88**]. See also Berman [**14**] for a sharper result in the Kähler-Einstein case in the presence of holomorphic vector fields.

Theorem 6.11. *If M admits a cscK metric in $c_1(L)$ and has discrete automorphism group, then (M, L) is K-stable.*

Example 6.12. Let $(M, L) = (\mathbf{CP}^1, \mathcal{O}(1))$, and embed $M \hookrightarrow \mathbf{CP}^2$ using the sections Z_0^2, Z_0Z_1, Z_1^2 of $\mathcal{O}(2)$ as a conic $xz - y^2 = 0$. Consider the $\mathbf{C}^*$-action $\lambda(t) \cdot (x, y, z) = (tx, t^{-1}y, z)$ as in Example 6.4. The central fiber is given by $xz = 0$, and the dual action on functions has weights $(-1, 1, 0)$ on (x, y, z). In order to compute the Donaldson-Futaki invariant, let us write

$$S = \mathbf{C}[x, y, z],$$

$I = (xz)$, and $R = S/I$. We have an exact sequence

$$0 \longrightarrow S \xrightarrow{\cdot xz} S \longrightarrow R \longrightarrow 0,$$

where the second map is multiplication by the generator xz of the ideal I. Let us write S_k and R_k for the degree k pieces of S and R. Let $D_k = \dim S_k$, and let W_k be the total weight of the $\mathbf{C}^*$-action on S_k. Similarly write $d_k = \dim R_k$ and w_k for the total weight of the action on R_k. From the exact sequence, for $k \geqslant 2$ we have

$$d_k = D_k - D_{k-2},$$

and since the weight of the action on xz is -1, we have

$$w_k = W_k - (W_{k-2} - D_{k-2}).$$

By symmetry of the weights on x, y we must have $W_k = 0$ for all k, and in addition

$$D_k = \binom{k+2}{2} = \frac{1}{2}(k^2 + 3k + 2).$$

It follows that

$$d_k = 2k + 1,$$
$$w_k = \frac{1}{2}k^2 - \frac{1}{2}k.$$

By the definition of the Donaldson-Futaki invariant, we have

$$F(\lambda) = \frac{1}{2} \cdot \frac{1}{2} + \frac{1}{2} = \frac{3}{4}.$$

In particular $F(\lambda) > 0$, so this test-configuration does not destabilize $\mathbf{CP}^1$. This is consistent with the fact that the Fubini-Study metric gives a cscK metric in $c_1(\mathcal{O}(1))$. More general calculations can be done similarly, except when the ideal I is not generated by a single polynomial. Then instead of the short exact sequence that we used, one would need to use a longer "free resolution" of the homogeneous coordinate ring.

Exercise 6.13. Let $X \subset \mathbf{CP}^2$ be the conic $xz - y^2 = 0$. Consider the $\mathbf{C}^*$-action $\lambda(t) \cdot (x, y, z) = (t^{-1}x, ty, z)$ (this is the opposite of Example 6.12). Find the Donaldson-Futaki invariant of the corresponding test-configuration.

6.4. Automorphisms and relative K-stability

In this section we will study how the notion of K-stability needs to be modified in the presence of automorphisms and in particular when we are interested in extremal metrics which do not have constant scalar curvature. In Section 6.1 we have seen that the scalar curvature can be seen as a moment map and that extremal metrics arise as critical points of the norm squared of this moment map. The discussion in Section 5.5 then indicates how we should modify the definition of K-stability for dealing with extremal metrics. In the next section we will give an example calculation of the resulting "relative K-stability" notion.

In Section 5.5 we have seen that in order to characterize orbits that contain critical points of $\|\mu\|^2$, we should test stability only in directions that commute with, and are orthogonal to, a maximal torus of automorphisms in a suitable sense.

Suppose as before that (M, L) is a polarized manifold, and let $T \subset \mathrm{Aut}(M, L)$ be a maximal torus of automorphisms. Given an embedding $M \hookrightarrow \mathbf{CP}^{N_r}$ using sections of L^r, we can realize T as a subgroup $T \subset GL(N_r + 1, \mathbf{C})$. Recall that a test-configuration for (M, L) of exponent r is given by a $\mathbf{C}^*$-subgroup of $GL(N_r + 1, \mathbf{C})$.

Definition 6.14. A test-configuration for (M, L) of exponent $r > 0$ is compatible with T if the corresponding $\mathbf{C}^*$-subgroup of $GL(N_r+1, \mathbf{C})$ commutes with T.

In analogy with Section 5.5 we are only interested in test-configurations that are "orthogonal" to T in a suitable sense. For this we need the following inner product, extending the norm in Equation (6.4). If two $\mathbf{C}^*$-subgroups λ, μ of $GL(N_r+1, \mathbf{C})$ commute and leave a subscheme $X \subset \mathbf{CP}^{N_r}$ invariant, then on the homogeneous coordinate ring $\bigoplus_{k\geqslant 0} R_k$ of X, the induced actions of λ, μ are generated by matrices A_k and B_k. We define the inner product $\langle \lambda, \mu \rangle$ by

$$(6.5)\quad \mathrm{Tr}\left[\left(A_k - \frac{\mathrm{Tr}(A_k)}{d_k}\mathrm{Id}\right)\left(B_k - \frac{\mathrm{Tr}(B_k)}{d_k}\mathrm{Id}\right)\right] = \langle \lambda, \mu \rangle k^{n+2} + O(k^{n+1}),$$

where $d_k = \dim R_k$. We will see in Section 7.4 that when X is smooth and λ, μ are generated by Hamiltonian vector fields, then this inner product is given by the L^2-product of the Hamiltonian functions. This is consistent with the fact that in viewing extremal metrics as critical points of the moment map squared, we are using the L^2-product on Hamiltonian functions. This inner product on holomorphic Hamiltonian vector fields was originally introduced by Futaki-Mabuchi [**56**] in their study of extremal metrics.

If a test-configuration for (M, L) is compatible with T, then T also acts on the central fiber of the test-configuration. We say that the test-configuration is orthogonal to T if the induced $\mathbf{C}^*$-action on the central fiber is orthogonal to every $\mathbf{C}^*$-subgroup of T. With this we have the following definition.

Definition 6.15. Let (M, L) be a polarized manifold and let $T \subset \mathrm{Aut}(M, L)$ be a maximal torus of automorphisms. The pair (M, L) is relatively K-semistable (relative to the torus T) if $F(\chi) \geqslant 0$ for all test-configurations χ for (M, L) which are orthogonal to T (and compatible with T). If in addition $F(\chi) > 0$ whenever $\|\chi\| > 0$, then (M, L) is relatively K-stable.

A further notion in the literature, relevant in the cscK case, is K-polystability. The pair (M, L) is K-polystable if it is relatively K-polystable and in addition the Futaki invariant of every vector field on M vanishes.

The conjecture analogous to Conjecture 6.9 is the following. One direction of this conjecture, the generalization of Theorem 6.11, has been obtained in [**102**].

Conjecture 6.16. *The manifold M admits an extremal metric in $c_1(L)$ if and only if (M, L) is relatively K-stable.*

In calculations, if we have a test-configuration χ that is compatible with a maximal torus T, then we can modify χ to be orthogonal to T by choosing an orthogonal basis of $\mathbf{C}^*$-actions $\lambda_1, \ldots, \lambda_l$ generating T and replacing χ by

$$\chi - \sum_{i=1}^{l} \frac{\langle \chi, \lambda_i \rangle}{\langle \lambda_i, \lambda_i \rangle} \lambda_i. \tag{6.6}$$

Since each λ_i commutes with χ, the central fiber will be unchanged; only the induced $\mathbf{C}^*$-action on it will be different.

6.5. Relative K-stability of a ruled surface

In this section we are going to work out the relative K-stability condition in the special case of the ruled surface that we studied in Section 4.4. Let Σ be a genus 2 curve and L a degree -1 line bundle on it. As before, X is the ruled surface $\mathbf{P}(L\oplus\mathcal{O})$ over Σ. In Section 4.4, following Tønnesen-Friedman [**116**], we constructed a family of extremal metrics on X which does not exhaust the entire Kähler cone. We will show that X is relatively K-unstable for the remaining polarizations, and so it does not admit an extremal metric. That X does not admit an extremal metric for these unstable polarizations was first shown in [**2**].

Since there are no non-zero holomorphic vector fields on Σ, a holomorphic vector field on X must preserve the fibers. Thus, the holomorphic vector fields on X are given by sections of $\mathrm{End}_0(\mathcal{O}\oplus\mathcal{M})$. Here End_0 means endomorphisms with trace zero. The vector field given by the matrix

$$\begin{pmatrix} -1 & 0 \\ 0 & 1 \end{pmatrix}$$

generates a $\mathbf{C}^*$-action β, which is a maximal torus of automorphisms (see Maruyama [**82**] for proofs).

The destabilizing test-configuration is an example of deformation to the normal cone of a subvariety, studied by Ross and Thomas [**92**]. We consider the polarization $\mathcal{L} = C + mS_\infty$ in the notation of Section 4.4; i.e. C is the divisor given by a fiber, S_∞ is the infinity section (ie. it satisfies $S_\infty^2 = 1$), and m is a positive constant. We denote by S_0 the zero section, so that $S_0 = S_\infty - C$. Note that β fixes S_0 and acts on the normal bundle of S_0 with weight 1. We will make no distinction between divisors and their associated line bundles, and we use the multiplicative and additive notations interchangeably, so for example $\mathcal{L}^k = kC + mkS_\infty$ for an integer k.

The deformation to the normal cone of S_0 can be most easily understood in terms of the "total space" of the test-configuration. While in Definition 6.6 we gave a very concrete definition of test-configurations, they can be thought of more geometrically, and that is how they were originally defined in [**112**] and [**44**]. Namely, if $X \subset \mathbf{CP}^N$ and $\chi : \mathbf{C}^* \to GL(N+1, \mathbf{C})$, then we can consider the (Zariski) closure of the set

$$\mathcal{X}^* = \{(\chi(t)\cdot x, t)\,|\, x\in X,\, t\in\mathbf{C}^*\} \subset \mathbf{CP}^N \times \mathbf{C}.$$

Denoting by $\mathcal{X}$ the closure, we have a projection map $\mathcal{X} \to \mathbf{C}$ such that the non-zero fibers are all isomorphic to X and the fiber over 0 is the central fiber X_0. The projective embedding is encoded in the restriction of $\mathcal{O}(1)$ to the family $\mathcal{X}$.

With this in mind, the deformation to the normal cone of S_0 is given by the blow-up

$$\mathcal{X} := \widetilde{X\times\mathbf{C}} \xrightarrow{\pi} X\times\mathbf{C}$$

along the subvariety $S_0 \times \{0\}$. Denote the exceptional divisor by E. For any rational $c \in (0, m)$ if k is sufficiently divisible, we have an ample line bundle $\mathcal{M}_c^k = k\pi^*\mathcal{L} - ckE$. For these values of c we therefore obtain a test-configuration $(\mathcal{X}, \mathcal{M}_c^k)$ with the $\mathbf{C}^*$-action induced by π from the product of the trivial action on X and the usual multiplication on $\mathbf{C}$. We denote the restriction of this $\mathbf{C}^*$-action to the central fiber $(X_0, \mathcal{L}_0)$ by χ.

We can view this test-configuration more concretely in terms of Definition 6.6. For a rational $c \in (0, m)$ we choose a large K such that cK is an integer and we can then embed $X \subset \mathbf{CP}^N$ using a basis of sections of $\mathcal{L}^K$. We define a $\mathbf{C}^*$-action on $\mathbf{CP}^N$ based on the order of vanishing along S_0. We can use the basis of sections to identify $\mathbf{CP}^N$ with $\mathbf{P}(H^0(X, \mathcal{L}^K))$. We now define a filtration on $H^0(X, \mathcal{L}^K)$ based on the order of vanishing along S_0:
(6.7)

$$F_0 \subset F_1 \subset \cdots \subset F_{cK} = H^0(X, \mathcal{L}^K),$$
$$F_i = \{s \in H^0(X, \mathcal{L}^K) \,|\, s \text{ vanishes along } S_0 \text{ to order at least } (cK - i)\}.$$

To define the $\mathbf{C}^*$-action, we choose an inner product on $H^0(X, \mathcal{L}^K)$ in order to turn the filtration into a direct sum decomposition

$$H^0(X, \mathcal{L}^K) = F_0 \oplus \frac{F_1}{F_0} \oplus \cdots \oplus \frac{F_{cK}}{F_{cK-1}},$$

and we define the action to have weight $-j$ on $F_j \big/ F_{j-1}$. Note that the resulting test-configuration is independent of the inner product that we choose (see Section 6.6).

We will need to modify this test-configuration χ by a multiple of the action β in order to make χ orthogonal to β, and we will need to compute the Futaki invariants. Note that β fixes the section S_0, and so χ commutes with β. This means that β induces an action on the central fiber of the test-configuration χ. For simplicity we will denote the actions on the central fiber by χ and β as well.

To calculate inner products and the Futaki invariants, we need to understand the actions of χ and β on the central fiber $(X_0, \mathcal{L}_0)$. In the following, k will always be a multiple of K, so in particular ck is an integer. According to [**92**] we have

$$H^0(X_0, \mathcal{L}_0^k) = H^0_X(k\mathcal{L} - ckS_0) \oplus \bigoplus_{j=1}^{ck} t^j \frac{H^0_X(k\mathcal{L} - (ck - j)S_0)}{H^0_X(k\mathcal{L} - (ck - j + 1)S_0)},$$

for k large, with t being the standard coordinate on $\mathbf{C}$. The $\mathbf{C}^*$-action χ has weight -1 on t. For the action β, we need to further decompose

$H^0_X(k\mathcal{L} - ckS_0)$ into weight spaces as follows:

$$H^0_X(k\mathcal{L} - ckS_0) = H^0_X(k\mathcal{L} - mkS_0) \oplus \bigoplus_{i=1}^{mk-ck} \frac{H^0_X(k\mathcal{L} - (mk-i)S_0)}{H^0_X(k\mathcal{L} - (mk-i+1)S_0)},$$

for k large. This holds because of the following cohomology vanishing lemma (see [**106**, Lemma 3.2.1]) from the Leray-Serre spectral sequence.

Lemma 6.17. *$H^1(X, kC + lS_\infty) = 0$ for $k \gg 0$ and $l \geqslant 0$.*

In sum we obtain the decomposition

$$\begin{aligned} H^0(X_0, \mathcal{L}_0^k) =& H^0_X(k\mathcal{L} - mkS_0) \oplus \bigoplus_{i=1}^{mk-ck} \frac{H^0_X(k\mathcal{L} - (mk-i)S_0)}{H^0_X(k\mathcal{L} - (mk-i+1)S_0)} \\ & \oplus \bigoplus_{j=1}^{ck} t^j \frac{H^0_X(k\mathcal{L} - (ck-j)S_0)}{H^0_X(k\mathcal{L} - (ck-j+1)S_0)}. \end{aligned} \tag{6.8}$$

As above, χ acts with weight -1 on t; that is, it acts with weight $-j$ on the summand of index j above. The action β acts on the term

$$\frac{H^0_X(k\mathcal{L} - lS_0)}{H^0_X(k\mathcal{L} - (l+1)S_0)}$$

with weight l, and the dimension of this space is $k + l - 1$ by the Riemann-Roch theorem. Let us write A_k, B_k for the infinitesimal generators of the actions χ and β on $H^0(X_0, L_0^k)$ and d_k for the dimension of this space. It is then straightforward to compute the following expansions:

$$\begin{aligned} d_k &= \frac{m^2+2m}{2}k^2 + \frac{2-m}{2}k + O(1), \\ \mathrm{Tr}(A_k) &= -\frac{c^3+3c^2}{6}k^3 + \frac{c^2-c}{2}k^2 + O(k), \\ \mathrm{Tr}(B_k) &= \frac{2m^3+3m^2}{6}k^3 + \frac{m}{2}k^2 + O(k), \\ \mathrm{Tr}(A_kB_k) &= -\frac{c^4+2c^3}{12}k^4 + O(k^3), \\ \mathrm{Tr}(B_kB_k) &= \frac{3m^4+4m^3}{12}k^4 + O(k^3). \end{aligned} \tag{6.9}$$

Following (6.6) we need to replace χ by

$$\widetilde{\chi} = \chi - \frac{\langle \chi, \beta \rangle}{\langle \beta, \beta \rangle}\beta$$

to make χ orthogonal to β, and then we need to compute $F(\widetilde{\chi})$. Using the formulas (6.9) and the definitions of the inner product and the Futaki invariant, we can compute

$$\begin{aligned}F(\widetilde{\chi}) &= F(\chi) - \frac{\langle \chi, \beta\rangle}{\langle \beta, \beta\rangle} F(\beta) \\ &= \frac{c(m-c)(m+2)}{4(m^2+6m+6)}\Big[(2m+2)c^2 - (m^2-4m-6)c + m^2 + 6m + 6\Big].\end{aligned}$$

If $F(\widetilde{\chi}) \leqslant 0$ for a rational c between 0 and m, then the variety is not relatively K-stable. Using this, we can check that $(X, \mathcal{L})$ is not relatively K-stable for $m \geq k_1 \cong 18.889$, where k_1 is the only positive real root of the quartic $m^4 - 16m^3 - 52m^2 - 48m - 12$. This should be compared with the calculation in Section 4.4, where we saw that X admits an extremal metric in $c_1(\mathcal{L})$ as long as $m < k_1$ with the same k_1. Note that the relative K-stability requires a certain polynomial to be positive for all $c \in (0, m) \cap \mathbf{Q}$, while the existence of an extremal metric requires a related polynomial to be positive for all $c \in (0, m)$. In this example these two conditions turn out to be the same.

Remark 6.18. Apostolov–Calderbank–Gauduchon–Tønnesen-Friedman [**2**] have generalized the above calculation together with the ODE existence result of Section 4.4 to a large family of examples. In particular they have found examples of ruled manifolds over products of Riemann surfaces, where the solution of the relevant ODE is positive on $(0, m) \cap \mathbf{Q}$ but vanishes at some irrational point. This shows that these examples do not admit extremal metrics, but if we only look at deformation to the normal cone as above, then the manifold appears to be stable. In other words, algebraically we only "see" the rational points in $(0, m)$. This leads to the expectation that we need to strengthen the notion of relative K-stability in a way that makes these ruled manifolds unstable, in order to have a chance for Conjecture 6.16 to be true.

6.6. Filtrations

In this section we will briefly describe a notion of stability which is a strengthening of K-stability. In particular it overcomes the issue raised in the previous section, namely the existence of manifolds which do not admit extremal metrics but which appear to be relatively K-stable. This notion was introduced in [**104**], motivated by work of Witt Nyström [**121**].

Recall that a test-configuration for a polarized manifold (M, L) is an embedding $M \hookrightarrow \mathbf{CP}^N$ using sections of L^r for some $r > 0$ together with a $\mathbf{C}^*$-action on $\mathbf{CP}^N$. For simplicity of notation, assume that $r = 1$. We want to define a filtration of the homogeneous coordinate ring of (M, L) from this data. If M is defined by the homogeneous ideal $I \subset \mathbf{C}[x_0, \ldots, x_N]$, then the homogeneous coordinate ring is

$$R = \mathbf{C}[x_0, \ldots, x_N]/I,$$

and we have a $\mathbf{C}^*$-action χ on $\mathbf{C}[x_0, \ldots, x_N]$. This $\mathbf{C}^*$-action defines a "weight filtration", $F_i \subset \mathbf{C}[x_0, \ldots, x_N]$, where

$$F_i = \operatorname{span}\{f \mid \chi(t) \cdot f = t^w f, \text{ for some } w \geqslant -i\}, \tag{6.10}$$

so F_i is the sum of the weight spaces with weights at least $-i$. This filtration in turn induces a filtration of R, defined by

$$F_i R = F_i \big/ I.$$

The construction of the flat limit in Definition 6.3 can be formulated more abstractly in terms of this filtration. In fact the homogeneous coordinate ring of the central fiber of the test-configuration turns out to be the associated graded ring

$$\operatorname{gr} R = \bigoplus_i F_i R \big/ F_{i-1} R,$$

with the product being defined through the multiplicative property

$$(F_i R)(F_j R) \subset F_{i+j} R$$

of the filtration. The $\mathbf{C}^*$-action χ on the central fiber acts on the i-th summand with weight $-i$, and it follows that we can compute the total weight w_k of the action on the degree k piece R_k from the filtration:

$$w_k = \sum_i -i \cdot (\dim F_i R_k - \dim F_{i-1} R_k). \tag{6.11}$$

Note that the sum only has finitely many non-zero terms. From this we can compute the Futaki invariant of the test-configuration.

In summary, this discussion shows that a test-configuration for (M, L) of exponent 1 can be encoded as a filtration of $H^0(M, L)$, and this in turn induces a filtration of the homogeneous coordinate ring. If we start with a test-configuration with exponent $r > 1$, then we can still apply this method to obtain a filtration of the subalgebra $\bigoplus_k H^0(M, L^{rk})$ of the homogeneous coordinate ring.

To obtain a stronger notion of stability we reverse this discussion and start with a filtration.

Definition 6.19. Let $R = \bigoplus R_k$ be the homogeneous coordinate ring of (M, L), where $R_k = H^0(M, L^k)$. A *filtration* of R is a collection of filtrations

$$\cdots \subset F_{i-1} R_k \subset F_i R_k \subset \cdots \subset R_k$$

satisfying the following conditions:

(1) The filtration is multiplicative; i.e.

$$(F_iR_k)(F_jR_l) \subset F_{i+j}R_{k+l} \quad \text{for all } i, j, k, l.$$

(2) For each k we have $F_mR_k = R_k$ for some m.

(3) There is a constant $C > 0$ such that $F_{-kC}R_k = \{0\}$ for all k.

This notion from [**18**], [**121**] is essentially equivalent, but slightly more general, than the notion defined in [**104**]. In the latter work we require $F_0R_k = \{0\}$ for $k > 0$, so no negative indices are used. Note that in [**18**] real numbers are allowed as indices, but in our setting integers seem more natural.

Suppose that we have such a filtration, which we will denote by χ. Then using property (2) above, for each r there is an induced filtration χ_r on R_r. By our previous discussion, χ_r can be thought of as a test-configuration for (M, L) of exponent r, and we would like to think of the original filtration χ as a kind of limit of the χ_r as $r \to \infty$. Because of this, it is natural to define the Futaki invariant $F(\chi)$ as a limit of the Futaki invariants $F(\chi_r)$, although there are also other natural choices. Similarly one can define the norm of χ as the limit of the norms of the χ_r:

$$F(\chi) = \liminf_{r\to\infty} F(\chi_r),$$
$$\|\chi\| = \lim_{r\to\infty} \|\chi_r\|,$$

where we used a lim inf in the first line because the limit is not known to exist in general. For these definitions to makes sense one needs to make sure that the Futaki invariant and norm are defined with a consistent scaling for test-configurations of different exponents. We refer the reader to [**104**] for details.

The main result in [**104**] is the following generalization of Theorem 6.11.

Theorem 6.20. *Suppose that M admits a cscK metric in $c_1(L)$ and has discrete automorphism group. Then $F(\chi) > 0$ for every filtration of the homogeneous coordinate ring of (M, L) which satisfies $\|\chi\| > 0$.*

In the presence of automorphisms, one needs to restrict the class of filtrations that are considered, analogously to the discussion in Section 6.4. If we have a maximal torus T of automorphisms of (M, L), then this induces a T-action on the homogeneous coordinate ring R and we say that a filtration χ of R is compatible with the torus action if each filtered piece F_iR is T-invariant. This ensures that the associated graded ring $\mathrm{gr}\, R$ inherits a T-action which commutes with the $\mathbf{C}^*$-action induced by χ. If β is any $\mathbf{C}^*$-subgroup of T, we can then define the inner product $\langle \chi, \beta \rangle$ just as in equation (6.5), with A_k and B_k denoting the infinitesimal generators of the actions χ, β on the k-th graded piece $\mathrm{gr}\, R_k$. We say that the filtration χ is orthogonal to T if it is compatible with T and its inner product with every $\mathbf{C}^*$-subgroup of T vanishes. With these preliminaries we define the following stronger notion of relative K-stability.

Definition 6.21. We say that the pair (M, L) is relatively $\widehat{K}$-stable if $F(\chi) > 0$ for all filtrations χ of the homogeneous coordinate ring of (M, L) which are orthogonal to a maximal torus of automorphisms of (M, L) and satisfy $\|\chi\| > 0$.

Remark 6.22. An even stronger notion of stability, $\overline{\mathrm{K}}$-stability, has been introduced by Donaldson [**50**]. This requires that not only is (M, L) K-stable, but also the blow-up $\mathrm{Bl}_p M$ should be K-stable for all $p \in M$ and all polarizations of the form $L - \varepsilon E$, where ε is a sufficiently small rational number and E is the exceptional divisor. Theorem 8.2 implies that if M admits a cscK metric in $c_1(L)$ and has no holomorphic vector fields, then (M, L) is $\overline{\mathrm{K}}$-stable. In addition the proof of Theorem 6.20 shows that $\overline{\mathrm{K}}$-stability implies $\widehat{\mathrm{K}}$-stability.

We now state a variant of the Yau-Tian-Donaldson conjecture.

Conjecture 6.23. *The manifold M admits an extremal metric in $c_1(L)$ if and only if (M, L) is relatively $\widehat{K}$-stable.*

One direction of this conjecture would follow from a suitable extension of Theorem 6.20 to the extremal case. It is likely that this can be done along the lines of [**102**] but it has not been worked out in the literature at this time.

To conclude this section, we will show how allowing filtrations overcomes the issue raised in Remark 6.18 at the end of the previous section. Following the notation in that section, recall that we defined the deformation to the normal cone of S_0 for any rational $c \in (0, m)$. Following [**121**] we can encode this as a filtration, so that if $R_k = H^0(X, \mathcal{L}^k)$, then for $i \geqslant 0$

$$F_i R_k = \{s \in R_k \mid s \text{ vanishes along } S_0 \text{ to order at least } (ck - i)\},$$

and $F_i R = \{0\}$ for $i < 0$. Note that this is an extension of the filtration in equation (6.7) to the whole homogeneous coordinate ring from R_K. This definition, however, makes sense even for irrational c. The crucial difference is that for irrational c the filtration no longer arises from a test-configuration. A calculation shows that in the examples of [**2**] referred to at the end of the last section, the relevant filtrations have vanishing Futaki invariants for exactly the values of $c \in (0, m)$ at which the corresponding ODE solutions vanish. In other words these examples are not potential counterexamples to Conjecture 6.23.

6.7. Toric varieties

In Section 4.5 we discussed the Kähler geometry of toric manifolds. In this section we take up the algebro-geometric point of view of toric varieties; in particular we will see how K-stability is related to the existence of cscK metrics, following Donaldson [**44**]. For a more in-depth introduction to toric varieties, see Fulton [**54**].

In order to make the closest contact with the point of view of Section 4.5 we will define toric varieties using their moment polytopes, through their homogeneous coordinate rings. Let $P \subset \mathbf{R}^n$ be a convex polytope with integral vertices, satisfying the Delzant condition (see Definition 4.35). Define $C(P) \subset \mathbf{R}^{n+1}$ to be the cone over $P \times \{1\}$ with vertex at the origin. Then $S_P = C(P) \cap \mathbf{Z}^{n+1}$ is a semigroup under addition, and we let $\mathbf{C}[S_P]$ be its semigroup algebra. This has a natural grading, with the degree k piece being spanned by the elements of S_P in $\mathbf{Z}^n \times \{k\}$. We can then define the projective variety

$$X_P = \operatorname{Proj} \mathbf{C}[S_P].$$

It turns out that if P satisfies the Delzant condition, then X_P is a smooth variety. In addition, writing X_P as Proj gives rise to an ample line bundle L_P on X_P whose global sections are the degree 1 elements in $\mathbf{C}[S_P]$.

We will now describe how this construction is related to the point of view in Section 4.5, in particular why P is the moment polytope. The variety X_P has an action of the torus $(S^1)^n$, induced by a torus action on the generators S_P of the homogeneous coordinate ring. This action has weight $p \in \mathbf{Z}^n$ on $(p, k) \in S_P$. We can find a moment map for this action by embedding $X_P \subset \mathbf{CP}^N$ and using the calculations in Section 5.1. We can embed X_P into projective space using the homogeneous functions corresponding to the points $p_0, \ldots, p_N \in P \cap \mathbf{Z}^n$. The fixed points in $\mathbf{CP}^N$ of the torus action are then the points with homogeneous coordinates

$$x_0 = [1 : 0 : \cdots : 0],\ x_1 = [0 : 1 : 0 : \cdots : 0], \ldots, x_N = [0 : \cdots : 0 : 1],$$

but not all of these points lie in X_P. In fact the points x_i that lie in X_P are precisely those for which p_i is a vertex of P. In addition, under suitable identifications the image of x_i under the moment map is the weight of the torus action at that point, i.e. p_i. The convexity theorem of Atiyah-Guillemin-Sternberg that we alluded to in Example 5.7 then implies that the image of the moment map is the convex hull of the vertices of P, so the image of the moment map is P itself.

Suppose that $f : P \to \mathbf{R}$ is a continuous convex function. We can define a filtration χ_f of the ring $R = \mathbf{C}[S_P]$ by letting

$$F_i R = \operatorname{span}\{(p, k) \in S_P \,:\, k f(k^{-1} p) \leqslant i\}. \tag{6.12}$$

The multiplicative property $(F_iR)(F_jR) \subset F_{i+j}R$ follows from the convexity of f. This filtration arises from a test-configuration if and only if f is rational piecewise linear, i.e. if it can be written as the maximum of a finite set of linear functions with rational coefficients. For instance if the coefficients are all integers, then the filtration χ_f is completely determined by the induced filtration on the degree 1 piece R_1. The corresponding test-configuration can be obtained by embedding $X_P \subset \mathbf{CP}^N$ using $p_0, \ldots, p_N \in P \cap \mathbf{Z}^n$ as above and acting on the i-th coordinate by $-f(p_i)$. It is useful to compare the filtrations (6.12) and (6.10) at this point. More generally when f is the maximum of rational linear functions where the denominators are all r, then the filtration corresponds to a test-configuration of exponent r, and we can reduce to the $r = 1$ case by replacing $\mathbf{Z}^n$ by $\frac{1}{r}\mathbf{Z}^n$.

Proposition 6.24. *The Donaldson-Futaki invariant of the test-configuration χ_f corresponding to a rational piecewise linear convex function f is*

$$F(\chi_f) = \frac{1}{2}\left(\int_{\partial P} f\, d\sigma - a\int_P f\, d\mu\right), \tag{6.13}$$

where $d\mu$ is the Lebesgue measure on P and $d\sigma$ is the measure on the boundary ∂P used in Lemma 4.37.

Proof. To simplify the notation, suppose that the test-configuration has exponent 1, i.e. f is the maximum of linear functions with integral coefficients. To use Definition 6.7 we need to compute the dimension d_k of the degree k piece of $\mathbf{C}[S_P]$ and the total weight w_k of the $\mathbf{C}^*$-action on it. The dimension d_k is simply the number $\#(P \cap \frac{1}{k}\mathbf{Z}^n)$ of elements of the lattice $\frac{1}{k}\mathbf{Z}^n$ in P. We have

$$\#\left(P \cap \frac{1}{k}\mathbf{Z}^n\right) = k^n \mathrm{Vol}(P, d\mu) + \frac{k^{n-1}}{2}\mathrm{Vol}(\partial P, d\sigma) + O(k^{n-2}). \tag{6.14}$$

As for the weight, using equation (6.11) we have

$$\begin{aligned} w_k &= -\sum_{(p,k)\in S_P} kf(k^{-1}p) \\ &= -\sum_{p\in P\cap\frac{1}{k}\mathbf{Z}^n} kf(p) \\ &= -k\left(k^n\int_P f\, d\mu + \frac{k^{n-1}}{2}\int_{\partial P} f\, d\sigma + O(k^{n-2})\right), \end{aligned}$$

where we used the result of Guillemin-Sternberg [**63**] in the last line (which also justifies (6.14) when setting $f = 1$). We can now use Definition 6.7 to obtain the required formula for $F(\chi_f)$. □

If f is not rational piecewise linear, then in the previous section we defined the Futaki invariant of χ_f to be

$$F(\chi_f) = \liminf_{r\to\infty} F(\chi_{f,r}),$$

where $\chi_{f,r}$ is the test-configuration of exponent r induced by the filtration on R_r. From definition (6.12) we can see that $\chi_{f,r}$ is the filtration induced by the rational piecewise linear convex function $f_r : P \to \mathbf{R}$ whose value at a point $p \in P \cap \frac{1}{r}\mathbf{Z}^n$ is

$$f_r(p) = \min\{r^{-1}i \,:\, i \in \mathbf{Z} \text{ with } f(p) \leqslant r^{-1}i\}.$$

In other words f_r is a rational approximation to f, and $f_r \to f$ uniformly as $r \to \infty$. It follows that we can take a limit in the formula (6.13) to obtain

$$F(\chi_f) = \liminf_{r\to\infty} F(\chi_{f_r}) = \frac{1}{2}\left(\int_{\partial P} f\, d\sigma - a\int_P f\, d\mu\right). \tag{6.15}$$

Since X_P always has a non-trivial torus of automorphisms, the natural stability notion is relative K-stability. The filtration χ_f is compatible with the maximal torus action, and any $\mathbf{C}^*$-subgroup β of the torus corresponds to a linear function h_β on P, defined by the weights of the action. By a similar argument to Proposition 6.24 we find that the inner product $\langle \chi_f, \beta\rangle$ is given by

$$\langle \chi_f, \beta\rangle = \int_P (f-\overline{f})(h_\beta - \overline{h}_\beta)\, d\mu,$$

where $\overline{f}, \overline{h}_\beta$ are averages, while the norm of χ_f is given by

$$\|\chi_f\|^2 = \int_P (f-\overline{f})^2\, d\mu.$$

It follows that if (X_P, L_P) is relatively K-stable, then for all non-zero rational piecewise linear convex functions $f : P \to \mathbf{R}$ which are L^2-orthogonal to all affine linear functions on P, we have

$$F(\chi_f) = \int_{\partial P} f\, d\sigma > 0. \tag{6.16}$$

It is more convenient to allow f which are not orthogonal to the affine linear functions. For this, let $A : P \to \mathbf{R}$ be an affine linear function defined by the condition that

$$\int_{\partial P} h\, d\sigma = \int_P Ah\, d\mu \quad \text{for all affine linear } h : P \to \mathbf{R}. \tag{6.17}$$

Then, if f is L^2-orthogonal to the affine linear functions and h is affine linear, we have

$$\begin{aligned}\int_{\partial P} f + h\, d\sigma &= \int_{\partial P} f\, d\sigma + \int_P Ah\, d\mu \\ &= \int_{\partial P} f\, d\sigma + \int_P A(f+h)\, d\mu.\end{aligned}$$

From this, together with (6.16), we have the following.

Proposition 6.25. *If (X_P, L_P) is relatively K-stable, then for all rational piecewise linear convex functions $f : P \to \mathbf{R}$ which are not affine linear, we have*

$$\int_{\partial P} f\, d\sigma - \int_P Af\, d\mu > 0, \tag{6.18}$$

where A is the unique affine linear function on P defined by (6.17).

This should be compared with the simple necessary condition (4.17) for the existence of an extremal metric that we obtained using integration by parts. In particular A is the scalar curvature of the extremal metric if it exists. In [**44**] an example is given of a polytope P such that the corresponding function A is constant and inequality (6.18) does not hold for suitable f, so (X_P, L_P) is not K-stable. In addition, note that by (6.15), if (X_P, L_P) is relatively $\widehat{\mathrm{K}}$-stable, then we can allow any continuous convex function $f : P \to \mathbf{R}$ rather than just the rational piecewise linear ones.

To conclude this section, we discuss what progress has been made on showing that relative K-stability of (X_P, L_P) implies the existence of an extremal metric on X_P. As in Chapter 3 the difficulty is to obtain a priori estimates for extremal metrics, so that one can use a suitable continuity method. More precisely, if A is an affine linear function on the polytope P and u is a symplectic potential on P satisfying $S(u) = A$, then we need to obtain estimates for u depending only on P and A. Since adding an affine linear function to u does not change the scalar curvature, we need to first normalize u in some way.

Definition 6.26. Let $x_0 \in P^\circ$ be a point in the interior of P. We say that a convex function $f : P \to \mathbf{R}$ is normalized if $f \geqslant 0$ and $f(x_0) = 0$.

In addition, the a priori estimates will need to depend on the stability condition since when (X_P, L_P) is not stable, then we cannot expect to be able to find an extremal metric. For this we define one more notion of stability which a priori is even stronger than $\widehat{\mathrm{K}}$-stability.

Definition 6.27. We say that (X_P, L_P) is uniformly stable if there is a $\lambda > 0$ with the following property. Fix a point $x_0 \in P^\circ$ in the interior of P.

Then for all normalized convex functions $f : P \to \mathbf{R}$, we have

$$\int_{\partial P} f\, d\sigma - \int Af\, d\mu \geqslant \lambda \int_{\partial P} f\, d\sigma,$$

where A is defined by (6.17).

Exercise 6.28. Let M be the blow-up of $\mathbf{CP}^2$ in one or two points. Show that the toric manifold (M, K_M^{-1}) is uniformly stable.

To see why this condition is useful, suppose that u is a symplectic potential on P such that $S(u) = A$. We can then apply the integration by parts formula of Lemma 4.37 to $g = u$ to obtain

$$\int_P n\, d\mu = \int_{\partial P} u\, d\sigma - \int_P Au\, d\mu.$$

If in addition we assume that u is normalized and that (X_P, L_P) is uniformly stable, then we obtain that

$$\int_{\partial P} u\, d\sigma \leqslant n\lambda^{-1}\mathrm{Vol}(P, d\mu).$$

This is a first a priori estimate on the solution u, and in the series of papers [**45**], [**48**], [**49**], Donaldson developed it into the following existence result.

Theorem 6.29. *Suppose that P is 2-dimensional, (X_P, L_P) is uniformly stable, and the affine linear function A is constant. Then X_P admits a cscK metric in $c_1(L_P)$.*

Together with the following result of Donaldson [**44**], this shows that the Yau-Tian-Donaldson conjecture for cscK metrics holds for toric surfaces.

Theorem 6.30. *Suppose that P is 2-dimensional and (X_P, L_P) is K-polystable (i.e. (X_P, L_P) is relatively K-stable, and the affine linear function A is constant). Then (X_P, L_P) is uniformly stable.*

Theorems 6.29 and 6.30 have been extended by Chen-Li-Sheng [**28**] and Wang-Zhou [**120**], respectively, to the case when the affine linear function A is not constant, thereby showing that the Yau-Tian-Donaldson conjecture for extremal metrics also holds for toric surfaces. In higher dimensions neither result is known, and it seems likely that Theorem 6.30 fails. It remains to be seen whether a stronger condition such as relative $\widehat{\mathrm{K}}$-stability is sufficient to imply uniform stability.

Chapter 7

The Bergman Kernel

In this chapter we discuss the asymptotic expansion of the Bergman kernel. This provides a crucial link between algebraic and differential geometry, and it is the basis of many results in the field. Our goal in this chapter will be to use it to prove Donaldson's theorem (Theorem 7.17), providing a lower bound on the Calabi functional in terms of the Futaki invariants of test-configurations [**46**]. A corollary of this result is that if an X admits a cscK metric in $c_1(L)$ for a line bundle L, then (X, L) is K-semistable.

7.1. The Bergman kernel

Let M be a compact complex manifold, and let L be a positive line bundle over M. Suppose that L is equipped with a Hermitian metric h which has positive curvature form $F(h)$. Let us define the Kähler form $\omega = \frac{1}{2\pi}F(h)$, so that $\omega \in c_1(L)$.

The Hermitian metric h induces a natural Hermitian metric on the space of holomorphic sections $H^0(M, L)$. For $s, t \in H^0(M, L)$ we define

$$\langle s, t\rangle_{L^2} = \int_M \langle s, t\rangle_h \frac{\omega^n}{n!}.$$

Definition 7.1. Choose an orthonormal basis $\{s_0, \ldots, s_N\}$ of $H^0(M, L)$. The Bergman kernel of the Hermitian metric h is the function

$$\begin{aligned} B_h : M &\to \mathbf{R} \\ x &\mapsto \sum_{i=0}^{N} |s_i(x)|_h^2. \end{aligned}$$

One can check that B_h is independent of the orthonormal basis chosen.

An alternative definition is given by the following.

Lemma 7.2. *For any* $x \in M$ *we have*

$$B_h(x) = \sup\{|s(x)|_h^2 \ : \ \|s\|_{L^2} = 1\}.$$

Proof. It is clear that $B_h(x) \geqslant |s(x)|_h^2$ for any s such that $\|s\|_{L^2} = 1$, by considering any orthonormal basis containing s.

For the converse inequality, write $E_x \subset H^0(M, L)$ for the space of sections vanishing at x. If $B_h(x) > 0$, then there must be a section which does not vanish at x, and so E_x has codimension 1. Let s be in the orthogonal complement of E_x such that $\|s\|_{L^2} = 1$. Then it follows from the definition that $B_h(x) = |s(x)|_h^2$ since every section orthogonal to s vanishes at x. □

The Bergman kernel has the following geometric interpretation.

Lemma 7.3. *Suppose that the map*

$$\begin{aligned} \varphi : M &\to \mathbf{CP}^N \\ x &\mapsto [s_0(x) : \cdots : s_N(x)] \end{aligned}$$

is defined on all of M, *where* $\{s_i\}$ *is an orthonormal basis of* $H^0(M, L)$. *Then*

$$\varphi^*\omega_{FS} = 2\pi\omega + \sqrt{-1}\partial\bar{\partial}\log B_h,$$

where ω_{FS} *is the Fubini-Study metric.*

Proof. On the subset of M where $s_0 \neq 0$, we have

$$\begin{aligned} \varphi^*\omega_{FS} &= \sqrt{-1}\partial\bar{\partial}\log\left(1 + \left|\frac{s_1}{s_0}\right|^2 + \cdots + \left|\frac{s_N}{s_0}\right|^2\right) \\ &= \sqrt{-1}\partial\bar{\partial}\log\left(1 + \frac{|s_1|_h^2}{|s_0|_h^2} + \cdots + \frac{|s_N|_h^2}{|s_0|_h^2}\right) \\ &= -\sqrt{-1}\partial\bar{\partial}\log|s_0|_h^2 + \sqrt{-1}\partial\bar{\partial}\log B_h \\ &= 2\pi\omega + \sqrt{-1}\partial\bar{\partial}\log B_h \end{aligned}$$

since $2\pi\omega$ is the curvature of h. The same argument works on the open sets where $s_i \neq 0$ for each i, and these cover M. □

The Hermitian metric h on L induces a metric h^k on L^k, and we get a corresponding Kähler form $k\omega$. Repeating the above construction with this metric, we obtain a function B_{h^k} on M. The key result is the asymptotic behavior of this function as $k \to \infty$.

Theorem 7.4. *As* $k \to \infty$, *we have*

$$B_{h^k} = 1 + \frac{S(\omega)}{4\pi}k^{-1} + O(k^{-2}), \tag{7.1}$$

where $S(\omega)$ is the scalar curvature of ω. More precisely, there are functions $a_0, a_1, \ldots$ on M such that $a_0 = 1$ and $a_1 = \frac{1}{4\pi}S(\omega)$, and for any integers $p, q \geqslant 0$ there is a constant C such that

$$\left\| B_{h^k} - \sum_{i=0}^{p} a_i k^{-i} \right\|_{C^q} \leqslant Ck^{-p-1}.$$

This theorem is due to Tian [**110**], Ruan [**93**], Zelditch [**124**], Lu [**78**], Catlin [**24**], and by now there is a large literature on it. In the next section we will only prove a simpler statement, giving the pointwise asymptotics (7.1). For now we look at some simple applications.

The original motivation of Tian was the following result, which implies that any Kähler metric in $c_1(L)$ can be approximated by "algebraic" metrics obtained as pullbacks of Fubini-Study metrics under projective embeddings. The result follows immediately from Lemma 7.3 and the previous theorem.

Corollary 7.5. *For large k, an orthonormal basis of $H^0(M, L^k)$ gives a map $\varphi_k : M \to \mathbf{CP}^{N_k}$, where $N_k + 1 = \dim H^0(M, L^k)$ and*

$$\frac{1}{k}\varphi_k^* \omega_{FS} - 2\pi\omega = O(k^{-2}) \quad \text{in } C^\infty.$$

Another application is the following special case of the Hirzebruch-Riemann-Roch theorem.

Corollary 7.6. *As $k \to \infty$, we have*

$$\dim H^0(M, L^k) = k^n \int_M \frac{\omega^n}{n!} + \frac{k^{n-1}}{4\pi} \int_M S(\omega)\frac{\omega^n}{n!} + O(k^{n-2}).$$

Proof. We integrate the expansion (7.1) over M, remembering that the $\{s_i\}$ form an orthonormal basis for $H^0(M, L^k)$. This means that

$$\begin{aligned} \dim H^0(M, L^k) &= \int_M \sum_{i=0}^{N_k} |s(x)|_h^2 \frac{(k\omega)^n}{n!} \\ &= \int_M B_{h^k} \frac{(k\omega)^n}{n!} \\ &= \int_M \left(1 + \frac{S(\omega)}{4\pi}k^{-1} + O(k^{-2})\right) \frac{k^n \omega^n}{n!} \\ &= k^n \int_M \frac{\omega^n}{n!} + \frac{k^{n-1}}{4\pi} \int_M S(\omega)\frac{\omega^n}{n!} + O(k^{n-2}). \end{aligned}$$

□

7.2. Proof of the asymptotic expansion

In this section we will prove the pointwise expansion (7.1) using the "peaked section" method of Tian [**110**], following the exposition of Donaldson [**50**].

Fix a point $x \in M$. The basic idea is to try to construct a holomorphic section σ of L^k such that $\|\sigma\|_{L^2} = 1$ and which is L^2-orthogonal to all holomorphic sections of L^k which vanish at x. If we could do this, then we would have $B_{h^k}(x) = |\sigma(x)|_h^2$. Although we cannot do this exactly, for large k it is possible to construct sections σ which are almost orthogonal to the sections which vanish at x. This is enough to calculate $B_{h^k}(x)$ up to an $O(k^{-2})$ error.

We can choose holomorphic coordinates w^i centered at x such that $2\pi\omega = \sqrt{-1}\partial\overline{\partial}\varphi$, where

$$\varphi(w) = |w|^2 - \frac{1}{4}R_{i\bar{j}k\bar{l}}w^i\overline{w}^j w^k \overline{w}^l + Q(w) + P(w).$$

Here $R_{i\bar{j}k\bar{l}}$ is the curvature tensor of $2\pi\omega$ at x, Q is a quintic polynomial, and $P(w) = O(|w|^6)$. Suppose the w^i are defined in a small neighborhood $B \subset M$ of x. For simplicity we can assume that $B = \{|w| < 1\}$. We can choose a holomorphic section s of L over B such that

$$|s|_h^2 = e^{-\varphi},$$

and for each k, we will use s^k to trivialize the bundle L^k over B.

Introduce coordinates $z^i = \sqrt{k}w^i$, and let $\Phi(z) = k\varphi(w)$. Then

$$\Phi(z) = |z|^2 - \frac{k^{-1}}{4}R_{i\bar{j}k\bar{l}}z^i\bar{z}^j z^k \bar{z}^l + k^{-3/2}Q(z) + kP(k^{-1/2}z), \tag{7.2}$$

and Φ is a Kähler potential for $2\pi k\omega$ in B. In terms of z we have $B = \{|z| < \sqrt{k}\}$. For large k, we can choose a cutoff function χ such that $\chi(z) = 1$ for $|z| < k^{1/5}$ and $\chi(z) = 0$ for $|z| > 2k^{1/5}$, and moreover $|\nabla\chi| < 1$. The reason for choosing $k^{1/5}$ is that on the ball $\{|z| < 2k^{1/5}\}$ we can make $\Phi(z)$ be arbitrarily close to $|z|^2$ by choosing k to be large enough. In particular for large k the metric $\sqrt{-1}\partial\overline{\partial}\Phi$ will be very close to the Euclidean metric.

The truncated function χs^k can be extended by zero outside B, and so it can be thought of as a global section σ_0 of L^k. It is not holomorphic, but

$$\overline{\partial}\sigma_0 = \overline{\partial}(\chi s^k) = (\overline{\partial}\chi)s^k$$

is supported in the annulus $\{k^{1/5} \leqslant |z| \leqslant k^{2/5}\}$, and

$$|\overline{\partial}\sigma_0|_{h^k}^2 \leqslant |s^k|_{h^k}^2$$

on this annulus. It follows that for large k

$$\|\overline{\partial}\sigma_0\|_{L^2}^2 \leqslant C\int_{k^{1/5}<|z|<2k^{1/5}} e^{-\frac{1}{2}|z|^2}\,dV = \varepsilon(k), \tag{7.3}$$

where by $\varepsilon(k)$ we mean a function of k that decays faster than any power of k. By construction $|\sigma_0(x)|^2_{h^k} = 1$.

The next task is to show that σ_0 can be perturbed to obtain a global holomorphic section of L^k. This uses the so-called Hörmander technique. The main point is the following.

Lemma 7.7. *Let $\Delta_{\overline{\partial}} = \overline{\partial}^*\overline{\partial} + \overline{\partial}\overline{\partial}^*$ be the $\overline{\partial}$-Laplacian on L^k-valued $(0,1)$-forms, where on L^k we use the metric h^k and on forms we use $2\pi k\omega$. If k is sufficiently large, then for any L^k-valued $(0,1)$-form α we have*

$$\langle \Delta_{\overline{\partial}}\alpha, \alpha\rangle_{L^2} \geqslant \frac{1}{2}\|\alpha\|^2_{L^2}. \tag{7.4}$$

Proof. This essentially follows from the Weitzenbock formula

$$\Delta_{\overline{\partial}} = \overline{\nabla}^*\overline{\nabla} + \mathrm{Ric} + F,$$

where Ric and F are endomorphisms obtained from the Ricci curvature of $k\omega$ and the curvature form of h^k, respectively. The point is that F is the identity, whereas as $k \to \infty$, Ric goes to zero. The details are as follows.

Let us write g for the metric $2\pi k\omega$, which is also the curvature of h^k. First note that

$$\langle \Delta_{\overline{\partial}}\alpha, \alpha\rangle_{L^2} = \|\overline{\partial}\alpha\|^2_{L^2} + \|\overline{\partial}^*\alpha\|^2_{L^2}.$$

In local coordinates let us write $\alpha = \alpha_{\bar{i}}d\bar{z}^i$, where the $\alpha_{\bar{i}}$ are sections of L^k. Let us work at a point x in normal coordinates for g. Then

$$\overline{\partial}\alpha = \sum_{j,k} \nabla_{\bar{j}}\alpha_{\bar{k}}d\bar{z}^j \wedge d\bar{z}^k = \sum_{j<k}(\nabla_{\bar{j}}\alpha_{\bar{k}} - \nabla_{\bar{k}}\alpha_{\bar{j}})d\bar{z}^j \wedge d\bar{z}^k,$$

where ∇ is the Chern connection on L^k coupled with the Levi-Civita connection on $(0,1)$-forms. Since the $d\bar{z}^j \wedge d\bar{z}^k$ form an orthonormal basis, we have

$$\begin{aligned}
|\overline{\partial}\alpha|^2 &= \sum_{j<k} |\nabla_{\bar{j}}\alpha_{\bar{k}}|^2 + |\nabla_{\bar{k}}\alpha_{\bar{j}}|^2 - \nabla_{\bar{j}}\alpha_{\bar{k}}\overline{\nabla_{\bar{k}}\alpha_{\bar{j}}} - \nabla_{\bar{k}}\alpha_{\bar{j}}\overline{\nabla_{\bar{j}}\alpha_{\bar{k}}} \\
&= \sum_{j,k} |\nabla_{\bar{j}}\alpha_{\bar{k}}|^2 - \nabla_{\bar{j}}\alpha_{\bar{k}}\overline{\nabla_{\bar{k}}\alpha_{\bar{j}}} \\
&= g^{p\bar{j}}g^{q\bar{k}}\nabla_{\bar{j}}\alpha_{\bar{k}}\overline{(\nabla_{\bar{p}}\alpha_{\bar{q}} - \nabla_{\bar{q}}\alpha_{\bar{p}})},
\end{aligned}$$

where we have used the summation convention in the last line and the metric h is implied in the pairing of L with $\overline{L}$. The last expression is coordinate invariant, so this is $|\overline{\partial}\alpha|^2$ even if we are not in normal coordinates. Note also that

$$\overline{\partial}^*\alpha = -g^{j\bar{k}}\nabla_j\alpha_{\bar{k}}.$$

We therefore have

$$\begin{aligned}\|\overline{\partial}\alpha\|_{L^2}^2 + \|\overline{\partial}^*\alpha\|_{L^2}^2 &= \int_M g^{p\bar{j}} g^{q\bar{k}} \big[\nabla_{\bar{j}}\alpha_{\bar{k}}\overline{(\nabla_{\bar{p}}\alpha_{\bar{q}} - \nabla_{\bar{q}}\alpha_{\bar{p}})} + \nabla_q\alpha_{\bar{k}}\overline{\nabla_j\alpha_{\bar{p}}}\,\big]\, dV \\ &= \|\overline{\nabla}\alpha\|_{L^2}^2 + \int_M g^{p\bar{j}} g^{q\bar{k}} \big[(\nabla_q\nabla_{\bar{j}}\alpha_{\bar{k}})\overline{\alpha_{\bar{p}}} - (\nabla_{\bar{j}}\nabla_q\alpha_{\bar{k}})\overline{\alpha_{\bar{p}}}\big]\, dV \\ &\geqslant \int_M g^{p\bar{j}} g^{q\bar{k}} \big[R^{\bar{m}}_{\bar{k}q\bar{j}}\alpha_{\bar{m}}\overline{\alpha_{\bar{p}}} + F_{q\bar{j}}\alpha_{\bar{k}}\overline{\alpha_{\bar{p}}}\big]\, dV \\ &= \int_M \big[g^{p\bar{j}} g^{q\bar{k}} R_{p\bar{k}}\alpha_{\bar{j}}\overline{\alpha_{\bar{q}}} + g^{p\bar{k}}\alpha_{\bar{k}}\overline{\alpha_{\bar{p}}}\big]\, dV.\end{aligned}$$

Since the Ricci form $R_{p\bar{k}}$ is invariant under scaling the metric ω, we have that $R_{p\bar{k}} = O(k^{-1}g)$ since g is the metric $2\pi k\omega$. For sufficiently large k, we will then have

$$\|\overline{\partial}\alpha\|_{L^2}^2 + \|\overline{\partial}^*\alpha\|_{L^2}^2 \geqslant \frac{1}{2}\|\alpha\|_{L^2}^2,$$

which is what we wanted to prove. □

It follows from this result that for large k the operator $\Delta_{\overline{\partial}}$ has trivial kernel, and since it is selfadjoint, it is invertible (by a result analogous to Theorem 2.13). Let us now return to our section σ_0. Define

$$\sigma = \sigma_0 - \overline{\partial}^*\Delta_{\overline{\partial}}^{-1}\overline{\partial}\sigma_0.$$

Using that $\Delta_{\overline{\partial}}$ commutes with $\overline{\partial}^*$, we can check that $\overline{\partial}^*\overline{\partial}\sigma = 0$, and so

$$\langle\overline{\partial}\sigma, \overline{\partial}\sigma\rangle = \langle\overline{\partial}^*\overline{\partial}\sigma, \sigma\rangle = 0.$$

It follows that $\overline{\partial}\sigma = 0$, so σ is a holomorphic section of L^k. A priori it could be the zero section; however, the estimates (7.3) and (7.4) imply that for large k

$$\begin{aligned}\|\sigma - \sigma_0\|_{L^2}^2 &= \|\overline{\partial}^*\Delta_{\overline{\partial}}^{-1}\overline{\partial}\sigma_0\|_{L^2}^2 \\ &= \langle\Delta_{\overline{\partial}}^{-1}\overline{\partial}\sigma_0, \overline{\partial}\sigma_0\rangle \\ &\leqslant 2\|\overline{\partial}\sigma_0\|_{L^2}^2 \\ &= \varepsilon(k).\end{aligned} \tag{7.5}$$

At the same time σ_0 is holomorphic on the ball $|z| < k^{1/5}$, so $\sigma - \sigma_0$ is also holomorphic on this ball. The L^2-bound and the estimate from Corollary 2.3 for harmonic functions imply that $|\sigma - \sigma_0|_{h^k}^2(x) = \varepsilon(k)$. This implies that

$$|\sigma(x)|_{h^k}^2 = 1 + \varepsilon(k),$$

so if k is large enough, σ does not vanish at x.

Next we want to show that σ is approximately orthogonal to every holomorphic section which vanishes at x.

Lemma 7.8. *There is a constant C independent of k such that*

$$|\langle \tau, \sigma \rangle_{L^2}| \leqslant Ck^{-1} \|\tau\|_{L^2}$$

for every holomorphic section $\tau \in H^0(M, L^k)$ vanishing at x.

Proof. Using the trivializing section s^k, we can think of τ as a holomorphic function of z which vanishes at $z = 0$. Then by the mean value theorem we have

$$\int_{|z|<k^{1/5}} \tau(z) e^{-|z|^2}\, dV = 0, \tag{7.6}$$

where dV is the Euclidean volume form. We need to see that this differs from $\langle \tau, \sigma \rangle_{L^2}$ by at most $Ck^{-1}\|\tau\|_{L^2}$. First of all, by (7.5), we have

$$\langle \tau, \sigma \rangle_{L^2} = \langle \tau, \sigma_0 \rangle_{L^2} + \varepsilon(k)\|\tau\|_{L^2}.$$

Also, recall that $\sigma_0 = \chi s^k$, with $\nabla\chi$ supported in the annulus $k^{1/5} < |z| < 2k^{1/5}$, on which we can assume that $\Phi(z) > \frac{1}{2}|z|^2$ and that $2\pi k\omega$ is uniformly equivalent to the Euclidean metric. It follows that

$$\langle \tau, \sigma \rangle_{L^2} = \varepsilon(k)\|\tau\|_{L^2} + \int_{|z|<k^{1/5}} \tau(z) e^{-\Phi(z)} \frac{1}{(2\pi)^n n!} (\sqrt{-1}\partial\bar{\partial}\Phi)^n.$$

Using expansion (7.2) combined with (7.6) for the leading term, we find that

$$\begin{aligned} |\langle \tau, \sigma \rangle_{L^2}| &\leqslant \varepsilon(k)\|\tau\|_{L^2} + Ck^{-1} \int_{|z|<k^{1/5}} |z|^4 |\tau(z)| e^{-|z|^2}\, dV \\ &\leqslant \varepsilon(k)\|\tau\|_{L^2} + Ck^{-1}\|\tau\|_{L^2}, \end{aligned}$$

where we used the Cauchy-Schwarz inequality in the last step, once again using the fact that on the set $\{|z| < k^{1/5}\}$ the metrics h and $2\pi k\omega$ are uniformly equivalent to $e^{-|z|^2}$ and the Euclidean metric, respectively. □

Finally we want to compute the L^2 norm of the section σ.

Lemma 7.9. *For large k we have*

$$\|\sigma\|^2_{L^2} = 1 - \frac{S_\omega(x)}{4\pi} k^{-1} + O(k^{-2}),$$

where S_ω is the scalar curvature of ω.

Proof. By the same arguments as in the previous lemma, up to an error of $\varepsilon(k)$, which we can ignore, it is enough to compute the L^2-norm of s^k on the ball $\{|z| < k^{1/5}\}$. I.e. we need to compute

$$\int_{|z|<k^{1/5}} e^{-\Phi(z)} \frac{1}{(2\pi)^n n!} (\sqrt{-1}\partial\bar{\partial}\Phi)^n. \tag{7.7}$$

From (7.2) we have the expansion

$$\frac{1}{n!}(\sqrt{-1}\partial\bar{\partial}\Phi)^n = \left[1 - k^{-1}R_{k\bar{l}}z^k\bar{z}^l + k^{-3/2}q(z) + O(k^{-2}|z|^4)\right]dV,$$

where $q(z)$ is a cubic polynomial in $z^i, \bar{z}^i$ and

$$dV = (\sqrt{-1})^n dz^1 \wedge d\bar{z}^1 \wedge \cdots \wedge dz^n \wedge d\bar{z}^n$$

is 2^n times the Euclidean volume form on $\mathbf{C}^n$. In addition we have

$$e^{-\Phi(z)} = e^{-|z|^2}\left(1 + \frac{k^{-1}}{4}R_{i\bar{j}k\bar{l}}z^i\bar{z}^jz^k\bar{z}^l - k^{-3/2}Q(z) + O(k^{-2}|z|^6)\right).$$

Extending integral (7.7) over all of $\mathbf{C}^n$ introduces an error of $\varepsilon(k)$, so we have

$$\begin{aligned}(2\pi)^n\|\sigma\|^2_{L^2} = &\int_{\mathbf{C}^n} e^{-|z|^2}\,dV \\ &+ k^{-1}\int_{\mathbf{C}^n} e^{-|z|^2}\left(\frac{1}{4}R_{i\bar{j}k\bar{l}}z^i\bar{z}^jz^k\bar{z}^l - R_{k\bar{l}}z^k\bar{z}^l\right)dV \\ &+ k^{-3/2}\int_{\mathbf{C}^n} e^{-|z|^2}\,[Q(z) - q(z)]\,dV \\ &+ O(k^{-2}).\end{aligned} \tag{7.8}$$

These integrals can be computed by integrating in each coordinate direction separately and using the following formulas. First we have the 1-dimensional integral

$$\int_{\mathbf{C}} e^{-t|z|^2}\sqrt{-1}dz \wedge d\bar{z} = \frac{2\pi}{t}.$$

Differentiating this with respect to t we obtain

$$\int_{\mathbf{C}} |z|^2 e^{-|z|^2}\sqrt{-1}dz \wedge d\bar{z} = 2\pi, \qquad \int_{\mathbf{C}} |z|^4 e^{-|z|^2}\sqrt{-1}dz \wedge d\bar{z} = 4\pi.$$

All other integrals where the number of z and $\bar{z}$ factors are not equal will vanish by the mean value theorem. This implies that

$$\begin{aligned}\int_{\mathbf{C}^n} e^{-|z|^2}R_{i\bar{j}k\bar{l}}z^i\bar{z}^jz^k\bar{z}^l\,dV &= \sum_{i=1}^n (2\pi)^{n-1}(4\pi)R_{i\bar{i}i\bar{i}} + \sum_{i\neq j}(2\pi)^n(R_{i\bar{i}j\bar{j}} + R_{i\bar{j}j\bar{i}}) \\ &= 2(2\pi)^n\sum_{i,j}R_{i\bar{i}j\bar{j}} \\ &= 2(2\pi)^n S_{2\pi\omega}(x),\end{aligned}$$

where $S_{2\pi\omega}$ is the scalar curvature of the metric $2\pi\omega$. Similarly,

$$\int_{\mathbf{C}^n} e^{-|z|^2}R_{k\bar{l}}z^k\bar{z}^l\,dV = \sum_k (2\pi)^n R_{k\bar{k}} = (2\pi)^n S_{2\pi\omega}(x).$$

The integral involving Q, q vanishes. From (7.8) we therefore obtain

$$\|\sigma\|_{L^2}^2 = 1 - \frac{S_{2\pi\omega}(x)}{2}k^{-1} + O(k^{-2}).$$

Finally note that $S_{2\pi\omega} = \frac{1}{2\pi}S_\omega$. □

We are now ready to complete the proof of the asymptotic expansion (7.1). Recall that $|\sigma(x)|_{h^k} = 1 + \varepsilon(k)$, so in particular $\sigma(x)$ does not vanish at x for large k. Let $E_x \subset H^0(M, L^k)$ be the space of sections vanishing at x, and let

$$\sigma = \eta + \tau \tag{7.9}$$

be the orthogonal decomposition of σ with $\eta \perp E_x$ and $\tau \in E_x$. Then

$$\|\eta\|_{L^2}^2 = \|\sigma\|_{L^2}^2 - \|\tau\|_{L^2}^2.$$

Since

$$\langle\tau, \tau\rangle_{L^2} = \langle\tau, \sigma\rangle_{L^2} \leqslant Ck^{-1}\|\tau\|_{L^2},$$

we have $\|\tau\|_{L^2} \leqslant Ck^{-1}$, from which it follows that

$$\|\eta\|_{L^2}^2 = \|\sigma\|_{L^2}^2 + O(k^{-2}) = 1 - \frac{S_\omega(x)}{4\pi}k^{-1} + O(k^{-2}). \tag{7.10}$$

Since η is orthogonal to every section vanishing at x and since $|\eta(x)|_{h^k}^2 = 1 + \varepsilon(k)$, the Bergman kernel at x is given by

$$B_{h^k}(x) = \frac{|\eta(x)|_{h^k}^2}{\|\eta\|_{L^2}^2} = 1 + \frac{S_\omega(x)}{4\pi}k^{-1} + O(k^{-2}).$$

This completes the proof of (7.1). In order to obtain stronger results, in particular the fact that the expansion holds in C^l-norms, not just pointwise, one needs to work harder, but it is possible to argue along similar lines (see Tian [**110**] or Ruan [**93**]). An alternative approach is to use Fourier analytic techniques, as in Zelditch [**124**].

Exercise 7.10. We have shown that if $L \to M$ is a positive line bundle and $p \in M$, then we can choose a large power L^k such that there is a holomorphic section σ of L^k which does not vanish at p. Use a similar technique to show that for large enough k there are holomorphic sections $\sigma_1, \ldots, \sigma_n$ such that the holomorphic functions $f_i = \frac{\sigma_i}{\sigma}$ defined near p are such that the ∂f_i span $\Omega_p^{1,0}M$. Using this, prove the Kodaira embedding theorem (Theorem 1.40).

Exercise 7.11. Let L be a positive line bundle over a compact Kähler manifold (M, ω) of dimension n. Choose a metric h on L with positive curvature, not related to ω. On $H^0(M, L^k)$, define the inner product

$$\langle s, t\rangle_{L^2} = \int_M \langle s, t\rangle_{h^k}\frac{\omega^n}{n!}.$$

Given any orthonormal basis $\{s_0, \ldots, s_{N_k}\}$ for this inner product, define the Bergman kernel

$$B_k(x) = \sum_{i=0}^{N_k} |s_i(x)|^2_{h^k}.$$

What are the first two terms in the asymptotic expansion of $B_k(x)$ as $k \to \infty$?

7.3. The equivariant Bergman kernel

In this section we will discuss an equivariant version of expansion (7.1) in the simplest case of a circle action. The set-up is the same as in Section 7.1 except that in addition we assume that we have an S^1-action on M preserving the complex structure and the metric ω and that the action of S^1 lifts to the total space of the line bundle L. In particular we will have S^1-actions on the spaces of sections $H^0(L^k)$.

Suppose that the S^1-action on M is generated by a vector field v with Hamiltonian H. Recall that this means

$$dH(w) = \omega(w, v)$$

for all vector fields w. We assume that the lifting of the S^1-action to L is related to the choice of Hamiltonian H in the following way: the action of the generator v on sections of L is given by

$$v \cdot s = \nabla_v s + 2\pi\sqrt{-1}Hs. \tag{7.11}$$

Note that this is the opposite of the action that we discussed in Remark 5.22, chosen to avoid minus signs below.

Let us denote by $2\pi\sqrt{-1}A$ the induced endomorphism of the vector space $H^0(M, L)$. Then A is a Hermitian matrix, and we define the equivariant Bergman kernel

$$B_h^{S_1} : M \to \mathbf{R}$$

$$x \mapsto \sum_{i=0}^{N} \langle As_i, s_i \rangle_h(x),$$

where $\{s_0, \ldots, s_N\}$ is an orthonormal basis of $H^0(M, L)$. Once again, this is independent of the orthonormal basis chosen.

As before, we are interested in the asymptotics of $B_{h^k}^{S^1}$ as $k \to \infty$, and we have the following.

Proposition 7.12. *As $k \to \infty$, we have the asymptotic expansion*

$$B_{h^k}^{S^1} = Hk + \frac{S(\omega)H}{4\pi} + O(k^{-1}).$$

Note that this actually follows from expansion (7.1) of the usual Bergman kernel:

Exercise 7.13. Prove Proposition 7.12 using Theorem 7.4, together with an explicit expression for the Hamiltonian function of the vector field v with respect to the pullback $\varphi_k^*\omega_{FS}$ of the Fubini-Study metric in the notation of Corollary 7.5.

A useful variant, however, is obtained as follows. For any smooth function $f : \mathbf{R} \to \mathbf{R}$ we define

$$B_{h^k}^{S^1,f}(x) = \sum_{i=0}^{N_k} \langle f(k^{-1}A_k)s_i, s_i\rangle_{h^k}(x), \tag{7.12}$$

where $2\pi\sqrt{-1}A_k$ is the action of v on $H^0(M, L^k)$ and $f(k^{-1}A_k)$ is defined using the spectral theorem. In particular if $\{s_i\}$ is an orthonormal basis of eigenvectors of A_k with corresponding eigenvalues λ_i, then

$$B_{h^k}^{S^1,f}(x) = \sum_0^{N_k} f(k^{-1}\lambda_i)|s_i(x)|_{h^k}^2.$$

For these "twisted" Bergman kernels, we have

Proposition 7.14. *As $k \to \infty$, we have the asymptotic expansion*

$$B_{h^k}^{S^1,f}(x) = f(H) + \frac{S(\omega)f(H)}{4\pi} + O(k^{-2}).$$

We will only show a weak form of this expansion. We can follow the strategy of the previous section; namely, given a point $x \in M$, if we have a section $\eta \in H^0(M, L^k)$ which is L^2-orthogonal to all holomorphic sections vanishing at x, then

$$B_{h^k}^{S^1,f}(x) = \frac{\langle f(k^{-1}A_k)\eta, \eta\rangle_{h^k}(x)}{\|\eta\|_{L^2}^2}. \tag{7.13}$$

We use the same section η that we have constructed in equation (7.9). We can approximate f with polynomials, and in view of equation (7.11) we then need to understand $(k^{-1}\nabla_v)^j\eta$. For this, note that with respect to the rescaled coordinates z^i in (7.2), the vector field v has coefficients of order $\sqrt{k}$, and also since η is holomorphic, we can use the $(1,0)$-part of v in the calculations. At the point x our section σ_0 gives a trivialization in which the connection 1-forms of the covariant derivative are given by

$$A_k = \partial_k(-\Phi(z)),$$

with $\Phi(z)$ as in (7.2). At the point x, corresponding to $z = 0$, we then have $A_k = 0$ and $\partial_i A_k = 0$. It follows that the section σ_0 satisfies $\nabla_i\sigma_0(x) =$

$\nabla_j \nabla_i \sigma_0(x) = 0$. All higher derivatives are also bounded independently of k. In terms of the vector field v this implies that

$$(k^{-1}\nabla_{v^{1,0}})^j \sigma_0(x) = O(k^{-3/2}),$$

for all $j > 0$.

Using the Schauder estimates and the fact that $\sigma - \sigma_0$ is holomorphic on the unit ball, we get that $(k^{-1}\nabla_{v^{1,0}})^j \sigma(x) = O(k^{-3/2})$ as well. The Schauder estimates applied to τ, together with $\|\tau\|_{L^2} \leqslant Ck^{-1}$, imply $(k^{-1/2}\nabla_{v^{1,0}})^j \tau(x) = O(k^{-1})$. It follows from these estimates that

$$(k^{-1}\nabla_v)^j \eta(x) = O(k^{-3/2}),$$

for all $j > 0$. The action of $k^{-1}A_k$ is given by

$$k^{-1}A_k \cdot s = \frac{1}{2\pi k\sqrt{-1}}\nabla_v s + Hs,$$

and so we get

$$(k^{-1}A_k)^j \cdot s(x) = H^j(x)s(x) + O(k^{-3/2}).$$

Using polynomial approximations to f we then obtain

$$f(k^{-1}A_k) \cdot s(x) = f(H(x))s(x) + O(k^{-3/2}).$$

We can now use this in equation (7.13) together with the formula (7.10) for the L^2-norm of η to get

$$B_{h^k}^{S^1,f}(x) = \frac{f(H(x))}{\|\eta\|_{L^2}^2} + O(k^{-3/2}) = f(H(x)) + \frac{S_\omega(x)f(H(x))}{4\pi} + O(k^{-3/2}).$$

As for the non-equivariant Bergman kernel expansion, the Fourier analytic method leads to more precise results, as can be seen in Zelditch [**125**] for instance.

7.4. The algebraic and geometric Futaki invariants

Suppose that (M, L) is a polarized variety with a $\mathbf{C}^*$-action λ (acting on both M and L). Let $\omega \in c_1(L)$ be a metric invariant under the S^1-subgroup. In this situation we can define the Futaki invariant differential-geometrically for the vector field generating the S^1-action as in equation (4.1) and also algebraically as in Definition 6.7. We will use the Bergman kernel expansion to show that these two definitions are the same up to a constant factor.

Let us write A_k for the infinitesimal generator of the $\mathbf{C}^*$-action λ on $H^0(M, L^k)$. By this we mean that the action is given by $t \mapsto t^{A_k}$, and so A_k has integral eigenvalues, which are the weights of the action. Recall that the Donaldson-Futaki invariant is defined by looking at the asymptotic behaviors

$$\dim H^0(M, L^k) = a_0 k^n + a_1 k^{n-1} + O(k^{n-2}),$$
$$\mathrm{Tr}(A_k) = b_0 k^{n+1} + b_1 k^n + O(k^{n-1})$$

for large k. Since we would like to make use of Proposition 7.14, we will use the action on $H^0(M, L^k)$ given by equation (7.11) rather than the dual action that we have used earlier. This means that the Donaldson-Futaki invariant is given by

$$F_{\mathrm{alg}}(\lambda) = b_1 - \frac{a_1}{a_0} b_0, \tag{7.14}$$

in order to match with the sign that we had previously.

On the differential geometric side, suppose that the vector field v generates the S^1-action on M, normalized so that the time 1 map generated by v is the identity. The fact that we have a lifting of the $\mathbf{C}^*$-action to L means that we have a Hamiltonian H for the vector field v with respect to ω (we will see this in the proof below). The differential geometric Futaki invariant of v is then

$$F_{\mathrm{dg}}(v) = \int_M H(S(\omega) - \hat{S}) \frac{\omega^n}{n!}.$$

We are using a slightly different definition from before, but the two only differ by a constant factor. We will prove the following.

Proposition 7.15. *In the situation above, we have* $F_{\mathrm{dg}}(v) = -4\pi F_{\mathrm{alg}}(\lambda)$.

Proof. We need to compute the coefficients a_0, a_1, b_0, b_1 differential-geometrically. For a_0, a_1 we have done this in Corollary 7.6, from which we have

$$a_0 = \int_M \frac{\omega^n}{n!}, \quad a_1 = \frac{1}{4\pi} \int_M S(\omega) \frac{\omega^n}{n!}.$$

To deal with b_0, b_1 we can integrate the expansion in Proposition 7.12 over M. If we choose an orthonormal basis of $H^0(M, L^k)$ consisting of eigenvectors of A, we find that

$$\int_M B^{S_1}_{h^k} \frac{(k\omega)^n}{n!} = \mathrm{Tr}(A_k).$$

From Proposition 7.12 we obtain

$$\mathrm{Tr}(A_k) = k^{n+1} \int_M H \frac{\omega^n}{n!} + \frac{k^n}{4\pi} \int_M S(\omega) H \frac{\omega^n}{n!} + O(k^{n-1}),$$

and so it follows that

$$b_0 = \int_M H \frac{\omega^n}{n!}, \quad b_1 = \frac{1}{4\pi} \int_M H S(\omega) \frac{\omega^n}{n!}.$$

Using these in the definition of $F_{\mathrm{alg}}(\lambda)$, we get

$$F_{\mathrm{alg}}(\lambda) = \frac{\hat{S}}{4\pi}\int_M H\,\frac{\omega^n}{n!} - \frac{1}{4\pi}\int_M HS(\omega)\,\frac{\omega^n}{n!} = \frac{1}{4\pi}F(v). \qquad \square$$

In a similar way, using Proposition 7.14, we can also relate the inner product and norm of $\mathbf{C}^*$-actions defined in equation (6.5) to L^2-products of corresponding Hamiltonian functions, which was the original definition due to Futaki-Mabuchi [**56**]. For this, suppose that we have two commuting $\mathbf{C}^*$-actions λ, μ on (M, L), and denote by A_k and B_k the infinitesimal generators of the actions on $H^0(M, L^k)$. On the one hand we have the inner product $\langle\lambda, \mu\rangle$ defined by the asymptotics

$$\mathrm{Tr}\left[\left(A_k - \frac{\mathrm{Tr}(A_k)}{d_k}\mathrm{Id}\right)\left(B_k - \frac{\mathrm{Tr}(B_k)}{d_k}\mathrm{Id}\right)\right] = \langle\lambda, \mu\rangle k^{n+2} + O(k^{n+1}),$$

while on the other hand we have two Hamiltonian functions H_λ, H_μ generating the $\mathbf{C}^*$-actions.

Proposition 7.16. *We have $\langle\lambda, \mu\rangle = \langle H_\lambda - \overline{H}_\lambda, H_\mu - \overline{H}_\mu\rangle_{L^2}$, where $\overline{H}_\lambda, \overline{H}_\mu$ denote averages.*

Proof. Using the polarization identity

$$\langle\lambda, \mu\rangle = \frac{1}{4}(\langle\lambda + \mu, \lambda + \mu\rangle - \langle\lambda - \mu, \lambda - \mu\rangle),$$

it is enough to focus on the case when $\lambda = \mu$.

The formula follows easily from the integrated form of Proposition 7.14. Namely, applying the result to the functions 1, x, and x^2, we get

$$\begin{aligned} d_k &= k^n \int_M \frac{\omega^n}{n!}, \\ \mathrm{Tr}(A_k) &= k^{n+1}\int_M H_\lambda\,\frac{\omega^n}{n!}, \\ \mathrm{Tr}(A_k^2) &= k^{n+2}\int_M H_\lambda^2\,\frac{\omega^n}{n!}, \end{aligned}$$

and the required result follows from these formulas. $\square$

7.5. Lower bounds on the Calabi functional

Our goal in this section is to explain the proof of Donaldson's theorem [**46**], giving lower bounds for the Calabi functional in terms of Futaki invariants of test-configurations. Rather than reproducing all of the details from [**46**], we will focus on the main ideas.

Suppose that (X, L) is a polarized manifold. Recall that a test-configuration for (X, L) (of exponent 1 for simplicity) consists of an embedding $X \subset$

$\mathbf{CP}^N$ using a basis of sections of L and a $\mathbf{C}^*$-action $\lambda : \mathbf{C}^* \hookrightarrow GL(N+1, \mathbf{C})$. The flat limit

$$X_0 = \lim_{t\to 0} \lambda(t) \cdot X$$

is a projective scheme fixed by the action λ. In Definition 6.7 we defined the Donaldson-Futaki invariant $F(X_0, \lambda)$ and the norm $\|\lambda\|$. We then have the following.

Theorem 7.17. *If $\omega \in c_1(L)$ is a Kähler metric on X, then*

$$\|\lambda\| \cdot \|S(\omega) - \hat{S}\|_{L^2} \geqslant -4\pi F(X_0, \lambda).$$

In particular if X admits a cscK metric in $c_1(L)$, then $F(X_0, \lambda) \geqslant 0$ for any test-configuration.

Note that by replacing L by L^r one can obtain similar statements for any test-configuration, not just those of exponent 1, so the conclusion $F(X_0, \lambda) \geqslant 0$ really holds for any test-configuration.

7.5.1. Using the Bergman kernel. Suppose that we have a projective manifold $V \subset \mathbf{CP}^N$ of dimension n. Define the matrix $M(V)$ to be

$$M(V)_{ij} = \int_V \frac{Z^i \overline{Z}^j}{|Z|^2} \frac{(\frac{1}{2\pi}\omega_{FS})^n}{n!},$$

and let $\underline{M}(V)$ be the trace free part of $M(V)$, i.e.

$$\underline{M}(V)_{ij} = M(V)_{ij} - \frac{\mathrm{Vol}(V)}{N+1}\delta_{ij}.$$

This is a moment map for the action of $SU(N+1)$ on the space of projective submanifolds of dimension n in $\mathbf{CP}^N$. The basic idea is that in some sense as $N \to \infty$, the moment map $\underline{M}$ approaches the infinite-dimensional moment map given by the scalar curvature, the link between the two being provided by the Bergman kernel expansion. We will now make this more precise.

Suppose that L is an ample line bundle on X, and let $\omega \in c_1(L)$.

Proposition 7.18. *There is a sequence of embeddings $M \to V_k \subset \mathbf{CP}^{N_k}$ using sections of L^k such that*

$$\|\underline{M}(V_k)\| \leqslant \frac{k^{n/2-1}}{4\pi} \|S(\omega) - \hat{S}\|_{L^2} + O(k^{n/2-2}),$$

where $\|M\|^2 = \mathrm{Tr}(M^2)$ for any Hermitian matrix M.

Proof. As in the construction of the Bergman kernel, let $\{s_i\}$ be an orthonormal basis of $H^0(X, L^k)$, where the inner product on sections is defined using a metric h on L whose curvature form is $2\pi\omega$. We let $V_k \subset \mathbf{CP}^{N_k}$ be the image of X under the embedding

$$\varphi_k : X \to \mathbf{CP}^{N_k},$$

given by this basis for large k. By applying a unitary transformation we can assume that $M(V_k)$ is diagonal, and so

$$M(V_k)_{ii} = \int_{V_k} \frac{|Z^i|^2}{|Z|^2} \frac{(\frac{1}{2\pi}\omega_{FS})^n}{n!} = \int_X |s_i|^2_{h^k} B_{h^k} \frac{(\frac{1}{2\pi}\varphi_k^*\omega_{FS})^n}{n!},$$

where B_{h^k} is the Bergman kernel. From Corollary 7.5 we know that

$$\left(\frac{1}{2\pi}\varphi_k^*\omega_{FS}\right)^n = (k\omega)^n\Big(1 + O(k^{-2})\Big)$$

and also

$$B_{h_k} = 1 + \frac{S(\omega)}{4\pi}k^{-1} + O(k^{-2}).$$

It follows that

$$\begin{aligned} M(V_k)_{ii} &= \int_X |s_i|^2_{h^k}\left(1 - \frac{S(\omega)}{4\pi}k^{-1}\right)\frac{(k\omega)^n}{n!} + O(k^{-2}) \\ &= 1 - \frac{k^{-1}}{4\pi}\int_X |s_i|^2_{h^k} S(\omega)\frac{(k\omega)^n}{n!} + O(k^{-2}). \end{aligned} \tag{7.15}$$

The rank of the matrix $M(V_k)$ is

$$N_k + 1 = \dim H^0(X, L^k) = \int_X \frac{(k\omega)^n}{n!} + O(k^{-1}),$$

and the trace is

$$\begin{aligned} \sum_{i=0}^{N_k} M(V_k)_{ii} &= N_k + 1 - \frac{k^{-1}}{4\pi}\int_X B_{h^k} S(\omega)\frac{(k\omega)^n}{n!} + O(k^{n-2}) \\ &= N_k + 1 - \frac{k^{-1}}{4\pi}\int_X S(\omega)\frac{(k\omega)^n}{n!} + O(k^{n-2}). \end{aligned}$$

It follows that

$$\frac{\mathrm{Tr}(M(V_k))}{N_k + 1} = 1 - \frac{k^{-1}}{4\pi}\hat{S} + O(k^{-2}),$$

and so using (7.15) the trace free part of $M(V_k)$ is

$$\underline{M}(V_k)_{ii} = \frac{k^{-1}}{4\pi}\int_X |s_i|^2_{h^k}(\hat{S} - S(\omega))\,\frac{(k\omega)^n}{n!} + O(k^{-2}),$$

where we also used that $\|s_i\|_{L^2} = 1$. Using the Cauchy-Schwarz inequality, we have

$$\begin{aligned} |\underline{M}(V_k)_{ii}|^2 &\leqslant \frac{k^{-2}}{16\pi^2}\int_X |s_i|^2\,\frac{(k\omega)^n}{n!}\int_X |s_i|^2(\hat{S} - S(\omega))^2\,\frac{(k\omega)^n}{n!} + O(k^{-3}) \\ &= \frac{k^{-2}}{16\pi^2}\int_X |s_i|^2(\hat{S} - S(\omega))^2\,\frac{(k\omega)^n}{n!} + O(k^{-3}), \end{aligned}$$

and so summing over the $N_k+1\sim k^n$ terms, we have

$$\begin{aligned}\|\underline{M}(V_k)\|^2 &\leqslant \frac{k^{-2}}{16\pi^2}\int_X B_{h^k}\,(\hat{S}-S(\omega))^2\,\frac{(k\omega)^n}{n!}+O(k^{n-3})\\ &= \frac{k^{n-2}}{16\pi^2}\int_X (\hat{S}-S(\omega))^2\,\frac{\omega^n}{n!}+O(k^{n-3}).\end{aligned}$$

Taking square roots gives the result we want. □

7.5.2. Lower bounds on $\|\underline{M}(V)\|$. Suppose that $V\subset \mathbf{CP}^N$ is a projective manifold, and let $\lambda:\mathbf{C}^*\hookrightarrow GL(N+1,\mathbf{C})$ be a one-parameter subgroup. We further require now that the image of the unit complex numbers $S^1\subset\mathbf{C}^*$ under λ lies in $U(N+1)$. We will give a lower bound for $\|\underline{M}(V)\|$ that is a finite-dimensional analog of Theorem 7.17.

For any $t\in\mathbf{C}^*$, let us write $V_t=\lambda(t)\cdot V$. Suppose that $\lambda(t)=t^A$ for a Hermitian matrix A with integer eigenvalues. Then the S^1-action on $\mathbf{CP}^N$ is induced by the skew Hermitian matrix $\sqrt{-1}A$ (we will use the convention (7.11) for the action on functions, which means that the formula for the Donaldson-Futaki invariant will be given by (7.14)). A Hamiltonian function for the vector field generating this circle action is then given by

$$h=\frac{A_{ij}Z^i\overline{Z}^j}{|Z|^2}.$$

Define the function

$$f(t)=\mathrm{Tr}(A\underline{M}(V^t))=\mathrm{Tr}(\underline{A}M(V^t)),$$

where $\underline{A}$ is the trace free part of A. Then

$$f(t)=\int_{V_t} h\,\frac{(\frac{1}{2\pi}\omega_{FS})^n}{n!}-\frac{\mathrm{Tr}(A)}{N+1}\mathrm{Vol(V)}.$$

The key point is that $f(t)$ is non-decreasing for $t\subset\mathbf{R}_{>0}$.

Lemma 7.19. *Restricting $f(t)$ to real numbers $t>0$, we have $f'(t)\geqslant 0$.*

Proof. This is essentially the calculation (5.3), using the fact that $\underline{M}$ is a moment map. Nevertheless we can check it directly. Let v_h be the vector field generating the S^1-action on $\mathbf{CP}^N$. Then

$$Jv_h=-\mathrm{grad}\,h.$$

Let us write $\Phi_s:\mathbf{CP}^N\to\mathbf{CP}^N$ for the one-parameter group of diffeomorphisms generated by Jv_h (this corresponds to approaching 0 along the positive real axis in $\mathbf{C}^*$). It is enough to compute the following derivative at $s = 0$:

$$\frac{d}{ds}\bigg|_{s=0} \int_{\Phi_s(V)} h \frac{\omega_{FS}^n}{n!} = \frac{d}{ds}\bigg|_{s=0} \int_V \Phi_s^*(h) \frac{(\Phi_s^*\omega_{FS})^n}{n!}$$
$$= \int_V (Jv_h)(h) \frac{\omega_{FS}^n}{n!} + \int_V h \frac{nL_{Jv_h}\omega_{FS} \wedge \omega_{FS}^{n-1}}{n!}.$$

Since $(-\mathrm{grad}\, h)(h) = -|\mathrm{grad}\, h|^2$, the first term is

$$\int_V -|\mathrm{grad}\, h|^2 \frac{\omega_{FS}^n}{n!}.$$

For the second term, recall from (6.3) that $L_{Jv_h}\omega_{FS} = -2\sqrt{-1}\partial\overline{\partial}h$. Integrating by parts we have

$$\int_V h\, 2n\sqrt{-1}\partial\overline{\partial}h \wedge \omega_{FS}^{n-1} = -\int_V 2n\partial h \wedge \overline{\partial}h \wedge \omega_{FS}^{n-1} = \int_V 2|\partial h|_V^2\, \omega_{FS}^n,$$

where we used Lemma 4.7 and we want to emphasize that $|\partial h|_V^2$ is the norm of only the part of ∂h which is tangential to V. In terms of the real gradient, $|\partial h|_V^2 = \frac{1}{2}|\mathrm{grad}\, h|_V^2$, where again only the tangential part is considered. It follows that

$$\frac{d}{ds}\bigg|_{s=0} \int_{\Phi_s(V)} h \frac{\omega_{FS}^n}{n!} = -\int_V |\mathrm{grad}\, h|_N^2 \frac{\omega_{FS}^n}{n!} \leqslant 0,$$

where $|\mathrm{grad}\, h|_N^2$ means that we are taking the norm of the normal component to V. Increasing t corresponds to flowing along $-Jv_h$, so the result that we want follows. □

Let us write $V_0 = \lim_{t\to 0} V_t$ for the flat limit. A crucial fact is that there is an algebraic cycle $|V_0|$ associated to V_0, which can be thought of as the union of the n-dimensional irreducible components of V_0, counted with multiplicities. In this way one can make sense of integrals over V_0, and if we define

$$FCh(A, V_0) = \int_{V_0} h \frac{(\frac{1}{2\pi}\omega_{FS})^n}{n!} - \frac{\mathrm{Tr}(A)}{N+1}\mathrm{Vol}(V),$$

then

$$\lim_{t\to 0} f(t) = FCh(A, V_0),$$

since the convergence $V_t \to |V_0|$ holds in the sense of currents. The monotonicity of $f(t)$ implies the following, which is the finite-dimensional analog of Theorem 7.17.

Proposition 7.20. *We have*

$$\|\underline{A}\| \cdot \|\underline{M}(V)\| \geqslant FCh(A, V_0).$$

Proof. The monotonicity of f implies that

$$\mathrm{Tr}(\underline{AM}(V)) = f(1) \geqslant \lim_{t\to 0} f(t) = FCh(A, V_0).$$

The result then follows from the Cauchy-Schwarz inequality. □

In order to relate this to the Futaki invariant, we need to be able to compute $FCh(A, V_0)$ algebro-geometrically. Recall from Section 6.3, that given the $\mathbf{C}^*$-action λ, there is an induced $\mathbf{C}^*$-action on the homogeneous coordinate ring

$$R = \mathbf{C}[x_0, \ldots, x_N]/I_0,$$

where I_0 is the homogeneous ideal corresponding to the flat limit V_0 (except that our convention is the opposite of that in Section 6.3). Let us write A_k for the generator of the $\mathbf{C}^*$-action on R_k, so the total weight of the action is $w_k = \mathrm{Tr}(A_k)$. Note that if $V \subset \mathbf{CP}^N$ is not contained in any hyperplane, then R_1 consists of all the linear polynomials, and $A_1 = A$. We need the following.

Lemma 7.21. *For large k we have*

$$\mathrm{Tr}(A_k) = k^{n+1} \int_{V_0} h \frac{(\frac{1}{2\pi}\omega_{FS})^n}{n!} + O(k^n). \tag{7.16}$$

If V_0 were smooth, then this would follow from integrating the expression in Proposition 7.12. In general, more involved arguments are required. One approach is to reduce the problem to the case when A has constant weights, and so h is a constant function. Then one needs to relate the volume of V_0 to the leading order term of its Hilbert polynomial (see Donaldson [**46**] for this approach). An alternative approach, following Wang [**119**], is to degenerate V_0 into a union of linear subspaces (with multiplicities) in such a way that the two sides of (7.16) remain unchanged in the limit and then check directly that the equation holds for linear subspaces.

7.5.3. Putting the pieces together.

We would now like to combine the results of the previous two subsections to complete the proof of Theorem 7.17. We start with a test-configuration λ for (X, L) of exponent 1 and a metric $\omega \in c_1(L)$. We would like to apply Proposition 7.20 to the sequence $V_k \subset \mathbf{CP}^{N_k}$ obtained in Proposition 7.18. For this we need to use λ to define $\mathbf{C}^*$-actions

$$\lambda_k : \mathbf{C}^* \hookrightarrow GL(N_k + 1, \mathbf{C})$$

in such a way that λ_k maps $S^1 \subset \mathbf{C}^*$ into $U(N_k + 1)$. A natural way to do this is to work with filtrations instead of $\mathbf{C}^*$-actions, as we did in Section 6.6.

Let us write $S = \mathbf{C}[x_0, \ldots, x_N]$ for the polynomial ring in $N+1 = \dim H^0(X, L)$ variables. The homogeneous coordinate ring of X is given by

$$R = S/I$$

for a homogeneous ideal I. As before, the $\mathbf{C}^*$-action λ induces a filtration (in fact even a grading) on S defined by

$$F_i S = \{\text{span of elements } f \in S \text{ with weights } \geqslant -i\},$$

and this descends to a filtration $F_i R$ on R such that

$$\cdots \subset F_i R \subset F_{i+1} R \subset \cdots .$$

We can further restrict the filtration to the degree k piece R_k for any k, and as we have seen before, all the data from the test-configuration that we need can be extracted from the filtration. The following is clear from the definitions.

Lemma 7.22. *We have the following:*

$$\begin{aligned} N_k + 1 = \dim R_k &= \sum_i (\dim F_i R_k - \dim F_{i+1} R_k), \\ \mathrm{Tr}(A_k) &= \sum_i i(\dim F_i R_k - \dim F_{i+1} R_k), \\ \mathrm{Tr}(A_k^2) &= \sum_i i^2(\dim F_i R_k - \dim F_{i+1} R_k). \end{aligned}$$

Note that there are only finitely many non-zero terms in each sum since for fixed k the filtration $F_i R_k$ must stabilize. Also, A_k denotes the generator of the $\mathbf{C}^*$-action on $H^0(X_0, L^k)$ as in the definition of the Donaldson-Futaki invariant, but with the opposite convention for the action in order to match with equation (7.11).

Now for any k, recall that the embedding $X \to V_k \subset \mathbf{CP}^{N_k}$ was given by an orthonormal basis, for a suitable choice of Hermitian metric on $R_k = H^0(X, L^k)$. Using this metric we can decompose R_k into an orthogonal direct sum

$$R_k = \cdots \oplus F_i R_k / F_{i-1} R_k \oplus F_{i+1} R_k / F_i R_k \oplus \cdots ,$$

where only finitely many terms are non-zero. We then define the $\mathbf{C}^*$-action λ_k to act with weight $-i$ on the summand $F_i R_k / F_{i-1} R_k$. In this way, S^1 will act by unitary transformations of R_k. Applying Proposition 7.20 to this action, we get

$$\|\underline{A}_k\| \cdot \|\underline{M}(V_k)\| \geqslant \int_{(V_k)_0} h_k \frac{(\frac{1}{2\pi}\omega_{FS})^n}{n!} - \frac{\mathrm{Tr}(A_k)}{N_k+1} \mathrm{Vol}(V_k), \tag{7.17}$$

with self-explanatory notation. From Lemma 7.21 we have for fixed k, as $l \to \infty$,

$$\mathrm{Tr}(A_{kl}) = l^{n+1} \int_{(V_k)_0} h_k \frac{(\frac{1}{2\pi}\omega_{FS})^n}{n!} + O(l^n).$$

Comparing this to the expansion (as $k \to \infty$)

$$\mathrm{Tr}(A_k) = b_0 k^{n+1} + O(k^n),$$

we see that

$$\int_{(V_k)_0} h_k \frac{(\frac{1}{2\pi}\omega_{FS})^n}{n!} = b_0 k^{n+1}.$$

From (7.17) we therefore have

$$\begin{aligned}(\mathrm{Tr}(\underline{A}_k^2))^{1/2} \cdot \|\underline{M}(V_k)\| &\geqslant b_0 k^{n+1} - \frac{b_0 k^{n+1} + b_1 k^n + O(k^{n-1})}{a_0 k^n + a_1 k^{n-1} + O(k^{n-2})} a_0 k^n \\ &= b_0 k^{n+1} - \left[b_0 k^{n+1} + b_1 k^n + O(k^{n-1})\right]\left[1 - \frac{a_1}{a_0} k^{-1} + O(k^{-2})\right] \\ &= -k^n \left(b_1 - \frac{a_1}{a_0} b_0\right) + O(k^{n-1}).\end{aligned}$$

Combining this with Proposition 7.18 and the definitions of the Donaldson-Futaki invariant $F(X_0, \lambda)$ and the norm $\|\lambda\|$, we have

$$\begin{aligned}\left(\|\lambda\| k^{n/2+1} + O(k^{n/2})\right) &\left(\frac{k^{n/2-1}}{4\pi} \|S(\omega) - \hat{S}\|_{L^2} + O(k^{n/2-2})\right) \\ &\geqslant -k^n F(X_0, \lambda) + O(k^{n-1}).\end{aligned}$$

Letting $k \to \infty$ we find

$$\|\lambda\| \cdot \|S(\omega) - \hat{S}\|_{L^2} \geqslant -4\pi F(X_0, \lambda),$$

which is the statement of Theorem 7.17.

The following corollary is immediate from Theorem 7.17.

Corollary 7.23. *Suppose that X admits a cscK metric $\omega \in c_1(L)$. Then (X, L) is K-semistable.*

In Section 8.6 we will see that together with a perturbation argument, this result can be used to show that (X, L) is in fact K-stable if X has no holomorphic vector fields and admits a cscK metric in $c_1(L)$.

Exercise 7.24. Use Theorem 7.17 to show that extremal metrics on projective manifolds minimize the Calabi functional in their Kähler class.

7.6. The partial C^0-estimate

In previous sections, we were concerned with the asymptotics of the Bergman kernel for a fixed background metric ω. An important insight due to Tian [**111**], [**112**] is that obtaining uniform control of the Bergman kernel for a family of metrics is the key to relating the existence of Kähler-Einstein metrics on Fano manifolds to algebro-geometric stability. In this section we will outline some of these ideas.

Recall from Section 3.5 that if M is a Fano manifold, i.e. $c_1(M) > 0$, then M does not always admit a Kähler-Einstein metric. In order to find Kähler-Einstein metrics, we try to find metrics $\omega_t \in c_1(M)$ which satisfy

$$\mathrm{Ric}(\omega_t) = t\omega_t + (1-t)\alpha, \tag{7.18}$$

where $\alpha \in c_1(M)$ is a fixed Kähler form. As we discussed in Section 3.5, there is a $T > 0$ such that a solution ω_t exists for $t \in [0, T)$, and the difficulty is in understanding what happens to ω_t as $t \to T$.

Letting K_M^{-1} be the anticanonical line bundle, for any metric $\omega \in c_1(K_M^{-1})$ we can choose a metric h on K_M^{-1} whose curvature form is ω. We will write $B_{\omega,k} = B_{h^k}$ for the Bergman kernel constructed using h^k.

Tian [**111**] conjectured the following, called the partial C^0-estimate.

Conjecture 7.25. *Given $\varepsilon > 0$, there are constants $k, c > 0$ depending on M, ε such that if ω satisfies $\mathrm{Ric}(\omega) > \varepsilon\omega$, then*

$$\inf_M B_{\omega,k} > c.$$

The importance of this conjecture stems from the formula in Lemma 7.3. In particular, by replacing k by a large multiple, we can assume that we have embeddings

$$\varphi_t : M \to \mathbf{CP}^N$$

using orthonormal bases of $H^0(K_M^{-k})$ with respect to the metrics h_t^k for $t \in [0, T)$, and if the partial C^0-estimate holds, then we will have a uniform lower bound on the $B_{\omega_t,k}$. One can show (see [**51**] for details) that under the assumption $\mathrm{Ric}(\omega) > \varepsilon\omega$ we have an upper bound for $B_{\omega,k}$, so we have a constant C such that

$$\sup_M |\log B_{\omega_t,k}| < C$$

for all $t \in [0, T)$. From Lemma 7.3 we have

$$\frac{1}{2\pi k}\varphi_t^*\omega_{FS} = \omega_t + \frac{1}{k}\sqrt{-1}\partial\bar{\partial}\log B_{\omega_t,k},$$

so the upshot is that on the level of C^0-bounds for the Kähler potentials, we can compare the metrics ω_t along the continuity method to certain "algebraic" metrics $\varphi_t^*\omega_{FS}$. This can be used to compare the values of an energy functional, such as the Mabuchi functional, at ω_t and $\varphi_t^*\omega_{FS}$, while at the same time an algebro-geometric stability condition can be used to control the energy functional on the space of algebraic metrics. For details of this approach, see Tian [**112**], [**114**] and Paul [**87**].

There has been much progress recently on Tian's conjecture (Conjecture 7.25). In proving the existence of Kähler-Einstein metrics on Fano surfaces with reductive automorphism group, Tian [**111**] showed that the partial C^0-estimate holds for a family of Kähler-Einstein surfaces. There was little progress in the higher-dimensional case until the work of Donaldson-Sun [**51**], which extended Tian's result to Kähler-Einstein manifolds in all dimensions. Soon afterwards, Chen-Donaldson-Sun [**32**], [**33**], [**34**] extended these results to families of metrics ω_t solving a variant of equation (7.18), where α is replaced by the current of integration $[D]$ along a suitable divisor $D \subset M$. Geometrically this amounts to ω_t satisfying the equation $\mathrm{Ric}(\omega_t) = t\omega_t$ on $M \setminus D$ and having conical singularities along D with cone angle $2\pi t$. This result was enough to relate the existence of a Kähler-Einstein metric on M to the K-stability of (M, K_M^{-1}) (see [**34**]). The techniques of Donaldson-Sun [**51**] were extended to the Kähler-Ricci flow by Tian-Zhang [**115**], giving a new proof of the existence of Kähler-Einstein metrics on K-stable Fano manifolds for dimension at most 3. In [**105**] we showed that the methods of Chen-Donaldson-Sun can be used to obtain the partial C^0-estimate for solutions of the usual continuity method (7.18), i.e. without using conical singularities. As for the general case of Conjecture 7.25, Jiang [**66**] showed, using the Ricci flow techniques of [**115**], that the conjecture holds in dimensions up to 3.

To conclude this section, we briefly explain the idea in Donaldson-Sun [**51**] which underpins all of these works. In the proof of the asymptotic expansion of Theorem 7.4, the basic idea is that the manifold (M, ω) is close to Euclidean space on a suitable scale. More precisely, there is a small radius $r > 0$ such that any ball $B_r \subset M$ of radius r is well approximated by the Euclidean unit ball B^{2n} when scaled to unit size. We could therefore glue a model section of the trivial bundle over B^{2n}, with exponential decay, onto the manifold M and use the Hörmander technique to perturb it to a holomorphic section.

When we have a family of metrics (M, ω_i), then to use a similar argument one needs some scale r at which each (M, ω_i) "looks" standard. The key difficulty is that in general the model is no longer just Euclidean space, as can be seen already in the case of complex surfaces where orbifold singularities can develop. The new input in [**51**] is the theory of Cheeger-Colding [**25**], [**26**], [**27**] on the structure of limit spaces of sequences of manifolds with lower Ricci curvature bounds. In the present situation these results imply, roughly speaking, that there is a suitable scale r at which a neighborhood of each point in (M, ω_i) is well approximated by a ball in the cone $C(Y)$ over a length space Y. At most points the cone is just $\mathbf{C}^n$, which is similar to the case when we have a fixed metric, but there are other possibilities. The crucial step is then to show that one has suitable model sections over these cones that can be glued onto the manifold using cutoff functions. In the end one does not obtain asymptotics as precise as in Theorem 7.4; however, one can still obtain the lower bound for the Bergman kernel required for Conjecture 7.25.

Chapter 8

CscK Metrics on Blow-ups

Suppose that M is a compact Kähler manifold with a cscK metric ω. In this section we will describe how one can construct cscK metrics on the blow-up of M at a point. We will only discuss the simplest setting in detail, namely when M has no holomorphic vector fields. We will briefly outline how more general results can be obtained, as in Arezzo-Pacard [**3**], [**4**], Arezzo-Pacard-Singer [**5**], and also [**108**].

8.1. The basic strategy

The technique used for constructing cscK metrics on the blow-up at a point is very general and is used in a wide variety of problems in geometric analysis. There are two main steps in the argument. Starting with a cscK metric ω on M and a point $p \in M$, one first constructs a family of metrics ω_ε on the blow-up $\mathrm{Bl}_p M$ depending on a small parameter $\varepsilon > 0$. The metrics ω_ε are obtained by modifying ω on a very small neighborhood of p, as indicated in Figure 8.1, and as $\varepsilon \to 0$, the metrics ω_ε converge to ω in a suitable sense (away from the point p). The second step is to perturb ω_ε in its Kähler class to obtain a cscK metric. This involves studying the linearization of the scalar curvature operator, and it will only be possible for sufficiently small ε.

When $n > 2$ we can get away with a fairly crude construction of the approximate solutions ω_ε. For the case when $n = 2$ we will construct a better approximate solution, while even more careful constructions are needed for the more refined results in [**4**], [**5**], [**108**]. The other step in the argument,

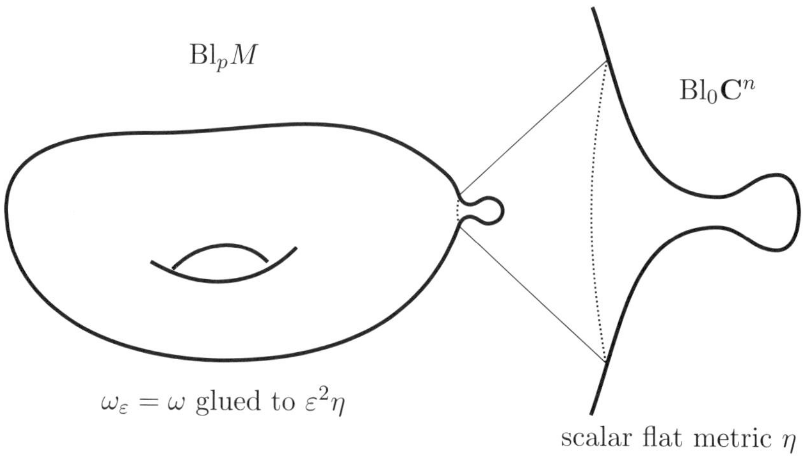

Figure 8.1. Constructing a metric ω_ε on Bl_pM.

namely studying the inverse of the linearized operator, is almost the same in each of these results.

8.1.1. Blow-ups. Let us recall briefly how to construct the blow-up of a Kähler manifold M at a point p. The blow-up $\mathrm{Bl}_0\mathbf{C}^n$ of $\mathbf{C}^n$ at the origin is simply the total space of the $\mathcal{O}(-1)$ bundle over $\mathbf{CP}^{n-1}$. Let us write

$$E \cong \mathbf{CP}^{n-1} \subset \mathrm{Bl}_0\mathbf{C}^n$$

for the zero section. There is a holomorphic map

$$\begin{aligned} \pi : \mathcal{O}(-1) &\to \mathbf{C}^n \\ \Big([z_0 : \cdots : z_n], (z_0, \ldots, z_n)\Big) &\mapsto (z_0, \ldots, z_n), \end{aligned}$$

where we recall that the fiber of $\mathcal{O}(-1)$ over the point $[z_0 : \cdots : z_n]$ is simply the line in $\mathbf{C}^{n+1}$ spanned by $(z_0, \ldots, z_n)$. The map π restricts to a biholomorphism

$$\pi : \mathrm{Bl}_0\mathbf{C}^n \setminus E \to \mathbf{C}^n \setminus \{0\}.$$

Suppose now that M is a complex manifold of dimension $n > 1$ and that $p \in M$. We can identify a neighborhood of p with a ball $B \subset \mathbf{C}^n$ such that p corresponds to the origin. The blow-up Bl_pM is then constructed by replacing $B \subset M$ by $\pi^{-1}(B) \subset \mathrm{Bl}_0\mathbf{C}^n$, using the biholomorphism

$$\pi : \pi^{-1}(B \setminus \{0\}) \to B \setminus \{0\}.$$

The result is a complex manifold Bl_pM, equipped with a holomorphic map

$$\pi : \mathrm{Bl}_pM \to M,$$

called the blow-down map. The preimage $E = \pi^{-1}(p)$ is a copy of $\mathbf{CP}^{n-1}$, and π restricts to a biholomorphism

$$\pi : \mathrm{Bl}_p M \setminus E \xrightarrow{\sim} M \setminus \{p\}.$$

Exercise 8.1. Let M be a compact complex manifold, and let $p \in M$. Show the following:

(a) Any holomorphic vector field on M which vanishes at p can be lifted to the blow-up $\mathrm{Bl}_p M$.

(b) Any holomorphic vector field on $\mathrm{Bl}_p M$ is obtained by lifting a holomorphic vector field on M that vanishes at p.

An application of the Mayer-Vietoris sequence shows that (if $n > 1$)

$$H^2(\mathrm{Bl}_p M, \mathbf{R}) \cong H^2(M, \mathbf{R}) \oplus \mathbf{R}[E],$$

where $[E]$ denotes the Poincaré dual of E. We will see that if ω is a Kähler metric on M, then for sufficiently small $\varepsilon > 0$, the class

$$\pi^*[\omega] - \varepsilon^2[E]$$

is a Kähler class on $\mathrm{Bl}_0 M$. Our goal is to prove the following theorem.

Theorem 8.2 (Arezzo-Pacard). *Suppose that M is a compact Kähler manifold with no holomorphic vector fields and that ω is a cscK metric on M. Then for any $p \in M$ the blow-up $\mathrm{Bl}_p M$ admits a cscK metric in the Kähler class*

$$\pi^*[\omega] - \varepsilon^2[E]$$

for sufficiently small $\varepsilon > 0$.

This theorem gives a way of constructing many new cscK manifolds. For instance we could take M to be a Kähler-Einstein manifold of dimension at least 2, given by Theorem 3.1. By Exercise 1.43, M does not admit holomorphic vector fields. We can then obtain new cscK metrics on the blow-up of M at any point, and we can even iterate the construction. Note that we have little understanding of what the metrics produced by Theorem 3.1 actually look like. In contrast we will see that the perturbation method giving Theorem 8.2 implies that the metrics we obtain on the blow-up $\mathrm{Bl}_p M$ are very close to our original metric on M away from the point p, while near p they are very close to scaled down versions of the the Burns-Simanca metric which we will study in the next section, as indicated by Figure 8.1.

8.1.2. The Burns-Simanca metric. A basic ingredient in constructing cscK metrics on blow-ups is a scalar flat, asymptotically flat metric on

$\mathrm{Bl}_0\mathbf{C}^n$. This metric was found by Burns and Simanca (see [**70**], [**95**]). For $n > 2$ the metric can be written as

$$\eta = \sqrt{-1}\partial\bar{\partial}\Big(|w|^2 + O(|w|^{4-2n})\Big) \tag{8.1}$$

as $|w| \to \infty$, in terms of the standard coordinates on $\mathrm{Bl}_0\mathbf{C}^n \setminus E \cong \mathbf{C}^n \setminus \{0\}$.

We will briefly describe the construction of the metric η here, using the methods we used in Section 4.4. We are trying to construct a metric on the total space of the $\mathcal{O}(-1)$-bundle over $\mathbf{CP}^{n-1}$. Choose a Hermitian metric h on $\mathcal{O}(-1)$ with curvature form $F(h) = -\omega_{FS}$. We will construct the metric in the form

$$\eta = \sqrt{-1}\partial\bar{\partial}f(s),$$

where f is a suitable strictly convex function and $s = \log|z|_h^2$ is the log of the fiberwise norm. As in Section 4.4 we can use coordinates z on $\mathbf{CP}^{n-1}$ and a fiberwise coordinate w, so that $|(z,w)|_h^2 = |w|^2 h(z)$. We can choose coordinates at a point such that $dh = 0$. Then

$$\eta = f'(s)p^*\omega_{FS} + f''(s)\frac{\sqrt{-1}dw \wedge d\bar{w}}{|w|^2},$$

where $p : \mathcal{O}(-1) \to \mathbf{CP}^{n-1}$ is the projection map. We get

$$\eta^n = \frac{f''(s)}{|w|^2}(f'(s))^{n-1}p^*\omega_{FS}^{n-1} \wedge \sqrt{-1}dw \wedge d\bar{w},$$

which is true at any point, not just where $dh = 0$. The Ricci form is therefore

$$\rho = -\sqrt{-1}\partial\bar{\partial}\log(f''(s)(f'(s))^{n-1}) + p^*(n\omega_{FS}),$$

using that $\mathrm{Ric}(\omega_{FS}) = n\omega_{FS}$. Taking the Legendre transform of f as in Section 4.4 we can rewrite this in terms of the function $\varphi(\tau) = f''(s)$, where $\tau = f'(s)$. We have

$$\begin{aligned}
\eta &= \tau p^*\omega_{FS} + \varphi(\tau)\frac{\sqrt{-1}dw \wedge d\bar{w}}{|w|^2},\\
\rho &= -\sqrt{-1}\partial\bar{\partial}\log(\varphi\tau^{n-1}) + np^*\omega_{FS}\\
&= \left(-\varphi' - \frac{(n-1)\varphi}{\tau} + n\right)p^*\omega_{FS} - \varphi\left(\varphi' + \frac{(n-1)\varphi}{\tau}\right)'\frac{\sqrt{-1}dw \wedge d\bar{w}}{|w|^2}.
\end{aligned}$$

Taking the trace, we have

$$S(\eta) = \frac{-1}{\tau^2}\Big[\tau^2\varphi'' + 2\tau(n-1)\varphi' + (n-1)(n-2)\varphi - \tau n(n-1)\Big].$$

To obtain a metric which on the zero section restricts to ω_{FS}, we need to find φ defined on $[1,\infty)$ such that

$$\varphi(1) = 0, \quad \varphi'(1) = 1.$$

The equation $S(\eta) = 0$ is equivalent to

$$\frac{d^2}{d\tau^2}\left[\tau^{n-1}\varphi\right] = \tau^{n-2}n(n-1). \tag{8.2}$$

Integrating this twice, using the boundary conditions, we get

$$\varphi(\tau) = \tau - (n-1)\tau^{2-n} + (n-2)\tau^{1-n}.$$

We need to change variables back to s to see the asymptotics of the Kähler potential in complex coordinates. Note that

$$\tau = \frac{d}{ds}f(s) = \varphi\frac{d}{d\tau}f(s),$$

so

$$\frac{d}{d\tau}f(s) = \tau\varphi^{-1} = (1 - (n-1)\tau^{1-n} + (n-2)\tau^{-n})^{-1}.$$

For large τ we have

$$\frac{d}{d\tau}f(s) = 1 + (n-1)\tau^{1-n} - (n-2)\tau^{-n} + O(\tau^{2-2n}),$$

and so up to changing f by a constant,

$$f(s) = \tau - \frac{n-1}{n-2}\tau^{2-n} + \frac{n-2}{n-1}\tau^{1-n} + O(\tau^{3-2n}).$$

We also have

$$\frac{ds}{d\tau} = \varphi^{-1} = \tau^{-1}(1 + (n-1)\tau^{1-n} - (n-2)\tau^{-n} + O(\tau^{2-2n}))$$

for large τ, so up to adding a constant to s (which corresponds to scaling the metric h), we have

$$\log|z|_h^2 = s = \log\tau + O(\tau^{1-n}).$$

Using this,

$$f(s) = |z|_h^2 - \frac{n-1}{n-2}|z|_h^{4-2n} + \frac{n-2}{n-1}|z|_h^{2-2n} + O(|z|^{6-4n}).$$

Now recall that under the biholomorphism $\mathcal{O}(-1)\setminus \mathbf{CP}^{n-1} \cong \mathbf{C}^n \setminus \{0\}$, the metric h is given by a multiple of the Euclidean metric

$$|(z_1,\ldots,z_n)|_h^2 = c(|z_1|^2 + \cdots + |z_n|^2).$$

This shows that the metric η is of the form given in equation (8.1).

When $n = 2$, then a similar calculation shows that

$$\eta = \sqrt{-1}\partial\bar{\partial}\Big(|w|^2 + \log|w|\Big).$$

Below, we will often not mention the $n = 2$ case separately since in the basic definitions, dealing with it is usually a simple modification. We will see, however, that when $n = 2$ there are some real difficulties when controlling the inverse of the linearized operator in Theorem 8.14, and also we will need to construct a better approximate solution than what we obtain in the next section. We will discuss these issues in Section 8.4.

8.1.3. The approximate solution. Let us suppose now that ω is a cscK metric on M and that we picked a point $p \in M$. In order to construct a metric on $\mathrm{Bl}_p M$ which has approximately constant scalar curvature, the idea is to replace the metric ω on a small neighborhood of p with a suitably scaled down copy of η. To do this we use cutoff functions to patch together the Kähler potentials.

Suppose that the z^i are normal coordinates centered at p, so that near p the metric ω is of the form

$$\omega = \sqrt{-1}\partial\overline{\partial}\Big(|z|^2 + \varphi_1(z)\Big),$$

where $\varphi_1(z) = O(|z|^4)$. For simplicity we can assume that the z^i are defined for $|z| < 1$. Fix a parameter ε, and let

$$r_\varepsilon = \varepsilon^{\frac{n-1}{n}}.$$

We will glue $\varepsilon^2\eta$ to ω on the annulus $B_{2r_\varepsilon} \setminus B_{r_\varepsilon}$. Under the change of variables $z = \varepsilon w$ we have

$$\varepsilon^2\eta = \sqrt{-1}\partial\overline{\partial}\Big(|z|^2 + \varepsilon^2\varphi_2(\varepsilon^{-1}z)\Big),$$

where $\varphi_2(z) = O(|z|^{4-2n})$.

Let us choose a smooth function $\gamma : \mathbf{R} \to [0,1]$ such that

$$\gamma(x) = \begin{cases} 1 \text{ if } x \geqslant 2, \\ 0 \text{ if } x \leqslant 1 \end{cases}$$

and define $\gamma_1(z) = \gamma(|z|/r_\varepsilon)$. Also, let $\gamma_2 = 1 - \gamma_1$. Define the metric ω_ε on $M \setminus \{p\}$ by letting

$$\omega_\varepsilon = \begin{cases} \omega \text{ on } M \setminus B_{2r_\varepsilon}, \\ \sqrt{-1}\partial\overline{\partial}\Big(|z|^2 + \gamma_1(z)\varphi_1(z) + \varepsilon^2\gamma_2(z)\varphi_2(\varepsilon^{-1}z)\Big) \text{ on } B_{2r_\varepsilon} \setminus B_{r_\varepsilon}, \\ \varepsilon^2\eta \text{ on } B_{r_\varepsilon} \setminus \{p\}. \end{cases}$$

The reason for our choice of $r_\varepsilon = \varepsilon^{(n-1)/n}$ is that, this way, on the annulus $B_{2r_\varepsilon} \setminus B_{r_\varepsilon}$ we have

$$\gamma_1(z)\varphi_1(z) + \varepsilon^2\gamma_2(z)\varphi_2(\varepsilon^{-1}z) = O(|z|^4).$$

The metric ω_ε is positive definite everywhere if ε is sufficiently small. It also naturally extends to a metric on $\mathrm{Bl}_p M$ which we will write as ω_ε as well. Since the volume of the exceptional divisor E with this metric is $\frac{\varepsilon^{2n-2}}{(n-1)!}$ and we have not changed ω outside a small ball, the Kähler class of ω_ε is $\pi^*[\omega] - \varepsilon^2[E]$.

8.1.4. The equation. Our goal is to perturb ω_ε into a cscK metric for sufficiently small ε. This means that we need to find a smooth function φ on Bl_pM such that $\omega_\varepsilon + \sqrt{-1}\partial\overline{\partial}\varphi$ is cscK. A small technical nuisance is caused by the fact that adding a constant to φ does not change the metric. One way to overcome this is to choose a point $q \in M$ outside the unit ball around p and try to solve the equation

$$S(\omega_\varepsilon + \sqrt{-1}\partial\overline{\partial}\varphi) - S(\omega) - \varphi(q) = 0. \tag{8.3}$$

This equation is no longer invariant under adding constants to φ.

We will solve the equation using the contraction mapping principle, and for this a crucial role is played by the linearization of the scalar curvature operator. At any metric ω this is given by Lemma 4.4 as

$$\begin{aligned} L_\omega(\varphi) := \frac{d}{dt}\bigg|_{t=0} S(\omega + t\sqrt{-1}\partial\overline{\partial}\varphi) &= -\mathcal{D}^*\mathcal{D}\varphi + g^{j\bar{k}}\partial_j S(\omega)\partial_{\bar{k}}\varphi \\ &= -\Delta^2\varphi - R_\omega^{j\bar{k}}\partial_j\partial_{\bar{k}}\varphi, \end{aligned}$$

where $R_\omega^{j\bar{k}}$ is the Ricci curvature of ω with the indices raised. An important observation is that if $S(\omega)$ is constant, then

$$L_\omega(\varphi) = -\mathcal{D}^*\mathcal{D}\varphi,$$

and so if M is compact, the kernel of $L_\omega(\varphi)$ coincides with the kernel of $\mathcal{D}$. If there are no non-zero holomorphic vector fields on M, then the kernel of $\mathcal{D}$ consists of only the constants, so in this case L_ω is an isomorphism when restricted to the L^2-orthogonal complement of the constants. Again one can remove the issue with the constant functions by considering the operator

$$\tilde{L}_\omega(\varphi) = L_\omega(\varphi) - \varphi(q),$$

where $q \in M$ is a point we fix in advance. It is then easy to check that if M is compact, ω is cscK, and M has no holomorphic vector fields, then

$$\tilde{L}_\omega : C^{k,\alpha}(M) \to C^{k-4,\alpha}(M) \tag{8.4}$$

is an isomorphism.

In order to solve equation (8.3), the most important step is to show that the linearization $\tilde{L}_{\omega_\varepsilon}$ is invertible and to obtain bounds on the norm of the inverse in suitable Banach spaces. It turns out that the right spaces to use are certain weighted Hölder spaces. In the next section we will discuss the basic theory of elliptic operators acting between weighted spaces.

8.2. Analysis in weighted spaces

8.2.1. The case of $\mathbf{R}^n \setminus \{0\}$. Many of the results that we need follow from a study of the Laplacian on $\mathbf{R}^n \setminus \{0\}$ in suitable weighted spaces. To define the weighted spaces, let $f : \mathbf{R}^n \setminus \{0\} \to \mathbf{R}$, and choose $\alpha \in (0,1)$, $\delta \in \mathbf{R}$, and a non-negative integer k. For $r > 0$, define

$$f_r : B_2 \setminus B_1 \to \mathbf{R}$$
$$f_r(x) = r^{-\delta} f(rx),$$

where B_r is the ball of radius r in $\mathbf{R}^n$. In other words f_r is the pullback of $r^{-\delta} f$ under the scaling map $B_2 \setminus B_1 \to B_{2r} \setminus B_r$. Then for any i we have

$$\nabla^i f_r(x) = r^{-\delta+i} \nabla^i f(rx).$$

Using this, we define the weighted norm by

$$\|f\|_{C^{k,\alpha}_\delta(\mathbf{R}^n \setminus \{0\})} = \sup_{r>0} \|f_r\|_{C^{k,\alpha}(B_2 \setminus B_1)} \tag{8.5}$$

in terms of the usual Hölder norms.

We say that $f \in C^{k,\alpha}_\delta(\mathbf{R}^n \setminus \{0\})$ if the weighted norm of f is finite. One can show that these weighted Hölder spaces are Banach spaces, and it follows from the definition that if $f \in C^{k,\alpha}_\delta$, then for $i \leqslant k$ we have $\nabla^i f \in C^{k-i,\alpha}_{\delta-i}$ and

$$\|\nabla^i f\|_{C^{k-i,\alpha}_{\delta-i}} \leqslant C \|f\|_{C^{k,\alpha}_\delta}$$

for some C independent of f. In particular for any δ the Laplacian defines a bounded linear map

$$\Delta_\delta : C^{k,\alpha}_\delta(\mathbf{R}^n \setminus \{0\}) \to C^{k-2,\alpha}_{\delta-2}(\mathbf{R}^n \setminus \{0\}).$$

Certain non-linear operators also define bounded maps between suitable weighted spaces, thanks to the boundedness of the multiplication maps

$$C^{k,\alpha}_\delta \times C^{k,\alpha}_{\delta'} \to C^{k,\alpha}_{\delta+\delta'}$$
$$(f, g) \mapsto fg.$$

The basic question is to determine the mapping properties of Δ_δ. The main result on this that we need is the following.

Theorem 8.3. *Suppose that $\delta \notin \mathbf{Z} \setminus (2-n, 0)$. Then the map*

$$\Delta_\delta : C^{k,\alpha}_\delta(\mathbf{R}^n \setminus \{0\}) \to C^{k-2,\alpha}_{\delta-2}(\mathbf{R}^n \setminus \{0\})$$

is an isomorphism.

Sketch of proof. The proof follows from studying explicit integral representations of the inverse of Δ. We will only give the proof in the easiest case, when $\delta \in (2-n, 0)$ and in particular $n > 2$. The general case (using weighted Sobolev spaces instead of Hölder spaces) can be found in Bartnik [**12**, Theorem 1.7].

Recall that the fundamental solution of the Laplacian is

$$G(x) = \frac{1}{n(n-2)c_n}|x|^{2-n},$$

where c_n is the volume of the unit ball in $\mathbf{R}^n$. If $u \in C^{k-2,\alpha}_{\delta-2}$, then we would like to define $f : \mathbf{R}^n \setminus \{0\} \to \mathbf{R}$ by

$$f(x) = \int_{\mathbf{R}^n\setminus\{0\}} G(x-y)u(y)\,dy.$$

To show that this is well-defined, we use that

$$|u(y)| \leqslant \|u\|_{C^{k-2,\alpha}_{\delta-2}}|y|^{\delta-2}.$$

By scaling and symmetry considerations it is enough to bound $|f(1)|$, which amounts to bounding the integral

$$\int_{\mathbf{R}^n\setminus\{0\}} \frac{|y|^{\delta-2}}{|1-y|^{n-2}}\,dy.$$

It follows that f is well-defined, $\Delta f = u$, and we have

$$|f(x)| \leqslant C\|u\|_{C^{k-2,\alpha}_{\delta-2}}|x|^{\delta}.$$

In order to obtain estimates for the derivatives of f, we can apply the Schauder estimates (Theorem 2.8) to the rescaled functions f_r in (8.5). This shows that Δ_δ is surjective. For more general weights δ one needs to use other integral kernels.

To see that Δ_δ is injective, suppose that $\Delta f = 0$, and expand f in spherical harmonics. In other words we write f in polar coordinates $f(r,\theta)$, where $\theta \in S^{n-1}$, and for each r we expand the function $\theta \mapsto f(r,\theta)$ in terms of eigenfunctions of the Laplacian on S^{n-1}. We have

$$f(r,\theta) = \sum_{i=1}^{\infty} a_i(r)\Psi_i(\theta),$$

where Ψ_i is an eigenfunction with eigenvalue λ_i. Using the formula for the Laplacian in polar coordinates, we have

$$\Delta f = \sum_{i=0}^{\infty}\left[a_i''(r) + \frac{n-1}{r}a_i'(r) + \frac{\lambda_i}{r^2}a_i(r)\right]\Psi_i(\theta).$$

It follows from this that if $\Delta f = 0$, then the $a_i(r)$ are solutions of the ODEs

$$a_i''(r) + \frac{2n-1}{r}a_i'(r) + \frac{\lambda_i}{r^2}a_i(r) = 0.$$

The solutions of this ODE are linear combinations of r^{s_1}, r^{s_2}, where

$$s_1, s_2 = n - 1 \pm \sqrt{(n-1)^2 - \lambda_i},$$

while the eigenvalues λ_i for $i \geqslant 0$ are given by

$$\lambda_i = -i(n-1+i).$$

The powers s_1, s_2, with λ_i ranging over all eigenvalues are called the *indicial roots* of Δ and they give all the possible growth rates of harmonic functions on $\mathbf{R}^n \setminus \{0\}$. From the above description we see that the indicial roots are

$$\{0,1,2,\ldots\} \cup \{2-n, 1-n, -n, \ldots\} = \mathbf{Z} \setminus (2-n, 0),$$

so our choice of δ means that Δ_δ is injective. □

Remark 8.4. From the expansion in terms of spherical harmonics above, it is clear that if instead of functions on $\mathbf{R}^n \setminus \{0\}$ we have $f : B_1 \setminus \{0\} \to \mathbf{R}$ or $g : \mathbf{R}^n \setminus B_1 \to \mathbf{R}$ and $\Delta f = \Delta g = 0$, then the growth rates of f and g at 0 and ∞, respectively, must be one of the indicial roots.

Exercise 8.5. Suppose that $u : \mathbf{R}^n \setminus \{0\} \to \mathbf{R}$ is in the weighted space $C^{k-2,\alpha}_{\delta-2}$ for some $k \geqslant 2$ and $\alpha \in (0,1)$. Suppose that $f : \mathbf{R}^n \setminus \{0\} \to \mathbf{R}$ satisfies

$$|f(x)| < C|x|^\delta$$

for some constant C and $\Delta f = u$. Show that then $f \in C^{k,\alpha}_\delta$.

The above result can be used to understand properties of the Laplacian in weighted spaces on manifolds such as $M_p = M \setminus \{p\}$ or on asymptotically flat manifolds such as $\mathrm{Bl}_0\mathbf{C}^n$. We can use our weighted spaces on $\mathbf{R}^{2n} \setminus \{0\}$ to define weighted spaces on such manifolds (n is the complex dimension) using cutoff functions. Consider M_p first. Let us use normal coordinates at p to identify a small geodesic ball, say B_1, around p, with the unit Euclidean ball $B_1 \subset \mathbf{R}^{2n}$. Let γ be a cutoff function on M, equal to 1 in $B_{1/2}$ and to 0 on $M \setminus B_1$. Then if $f : M_p \to \mathbf{R}$, we can think of γf, extended by zero, as a function on $\mathbf{R}^{2n} \setminus \{0\}$. We can therefore define the weighted norm

$$\|f\|_{C^{k,\alpha}_\delta(M_p)} = \|f\|_{C^{k,\alpha}(M\setminus B_{1/2})} + \|\gamma f\|_{C^{k,\alpha}_\delta(\mathbf{R}^{2n}\setminus\{0\})}.$$

Similarly on $\mathrm{Bl}_0\mathbf{C}^n$, let γ be a cutoff function which equals 1 outside B_2 and which equals 0 in B_1. Then for any $f : \mathrm{Bl}_0\mathbf{C}^n \to \mathbf{R}$ the function γf can be thought of as a function on $\mathbf{R}^{2n} \setminus \{0\}$. We can then define

$$\|f\|_{C^{k,\alpha}_\delta(\mathrm{Bl}_0\mathbf{C}^n)} = \|f\|_{C^{k,\alpha}(B_2)} + \|\gamma f\|_{C^{k,\alpha}_\delta(\mathrm{Bl}_0\mathbf{C}^n\setminus B_1)},$$

where B_1, B_2 are subsets of $\mathrm{Bl}_0\mathbf{C}^n$.

A crucial point is that we can compare the Laplacian Δ on M near p with the Euclidean Laplacian Δ_{Euc} on $B_1 \subset \mathbf{R}^{2n}$ in these weighted spaces, owing to the fact that the metric is flat to order 1 at p in normal coordinates. It follows that if $f \in C^{k,\alpha}_\delta(M_p)$, then

$$\Delta f - \Delta_{\mathrm{Euc}} f \in C^{k-2,\alpha}_\delta \quad \text{rather than } C^{k-2,\alpha}_{\delta-2},$$

and an analogous result holds on $\mathrm{Bl}_0\mathbf{C}^n$ when using a metric that is asymptotically flat. Using this, many results about the Euclidean Laplacian can be generalized to more general Laplacian operators.

Although we will not have to use this, we mention a central result in the theory. Note that on M_p and $\mathrm{Bl}_0\mathbf{C}^n$ we can no longer expect an isomorphism result as in Theorem 8.3; however, it turns out that except when δ is an indicial root, the Laplacian Δ_δ defines a Fredholm map. References for such results are Lockhart-McOwen [**77**], Pacard [**86**], or Bartnik [**12**].

Theorem 8.6. *If δ is not an indicial root, then the maps*

$$\Delta_\delta : C^{k,\alpha}_\delta(M_p) \to C^{k-2,\alpha}_{\delta-2}(M_p),$$
$$\Delta_\delta : C^{k,\alpha}_\delta(\mathrm{Bl}_0\mathbf{C}^n) \to C^{k-2,\alpha}_{\delta-2}(\mathrm{Bl}_0\mathbf{C}^n)$$

are Fredholm operators. Moreover

$$\mathrm{Im}(\Delta_\delta) = (\mathrm{Ker}(\Delta_{2-n-\delta}))^\perp,$$

where the orthogonal complement is taken with respect to the L^2-product.

In contrast to this, one can check that when $\delta = 0$, the image of Δ_δ is not closed, so the operator is not Fredholm:

Exercise 8.7. Consider the Laplacian acting between weighted spaces on M_p:

$$\Delta_\delta : C^{k,\alpha}_\delta(M_p) \to C^{k-2,\alpha}_{\delta-2}(M_p).$$

Show that if $\delta = 0$, then the image of Δ_δ is not closed, by considering the Laplacian of functions that behave like $\log\log|x|$ for x sufficiently close to p.

8.2.2. The linearized operator. Instead of the Laplacian, we are interested in the operator

$$L : \varphi \mapsto -\Delta^2\varphi - R^{j\bar{k}}\partial_j\partial_{\bar{k}}\varphi,$$

either with respect to the metric η on $\mathrm{Bl}_0\mathbf{C}^n$ or with respect to ω on M_p. The mapping properties of this operator can be deduced from those of Δ^2 acting on $\mathbf{R}^{2n}\setminus\{0\}$, using the fact that η is asymptotically flat at infinity and ω is flat up to first order at p. The mapping properties of Δ^2 on $\mathbf{R}^{2n}\setminus\{0\}$ can be deduced from those of Δ. In particular the indicial roots of Δ^2 on $\mathbf{R}^{2n}$ are $\mathbf{Z}\setminus(4-2n,0)$:

Exercise 8.8. Use Theorem 8.3 to show that for $\delta \notin \mathbf{Z}\setminus(4-n,0)$, the map

$$\Delta^2_\delta : C^{k,\alpha}_\delta(\mathbf{R}^n\setminus\{0\}) \to C^{k-4,\alpha}_{\delta-4}(\mathbf{R}^n\setminus\{0\})$$

is an isomorphism.

The following two results will be crucial.

Proposition 8.9. *If $\delta < 0$, then the operator*

$$L_\eta : C^{4,\alpha}_\delta(\mathrm{Bl}_0\mathbf{C}^n) \to C^{0,\alpha}_{\delta-4}(\mathrm{Bl}_0\mathbf{C}^n)$$

has trivial kernel.

Proof. We can assume that in fact $\delta \in (-1,0)$. We have $L_\eta = -\mathcal{D}^*_\eta\mathcal{D}_\eta$, so suppose that $\mathcal{D}^*_\eta\mathcal{D}_\eta\varphi = 0$ with $\varphi \in C^{4,\alpha}_\delta(\mathrm{Bl}_0\mathbf{C}^n)$ and $\delta < 0$. We first want to obtain better decay for φ at infinity, using our knowledge of the growth rates of biharmonic functions on $\mathbf{R}^{2n}$ at infinity (using Remark 8.4 and Exercise 8.8). Our formula for the metric η implies that as $|z| \to \infty$, we have

$$\eta_{j\bar{k}} = \delta_{j\bar{k}} + O(|z|^{2-2n}).$$

Let γ be a cutoff function supported in $\mathrm{Bl}_0\mathbf{C}^n \setminus B_1$, equal to 1 outside B_2. We can then think of $\gamma\varphi$ as a function on $\mathbf{C}^n = \mathbf{R}^{2n}$, and we can compare $\mathcal{D}^*\mathcal{D}$ with the Euclidean operator Δ^2. We obtain

$$\Delta^2(\gamma\varphi) \in C^{0,\alpha}_{\delta-2-2n}(\mathbf{C}^n \setminus \{0\}).$$

Exercise 8.8 implies that we can find a $\psi \in C^{4,\alpha}_{\delta+2-2n}(\mathbf{C}^n \setminus \{0\})$ such that

$$\Delta^2(\psi) = \Delta^2(\gamma\varphi).$$

It follows that $\psi - \gamma\varphi$ is a biharmonic function and it decays at infinity. Since there are no indicial roots in $(4-2n, 0)$, we must have

$$\psi - \gamma\varphi \in C^{4,\alpha}_{4-2n}(\mathbf{C}^n \setminus B_1),$$

at least when $n > 2$, while for $n = 2$ we already have better decay than this by assumption. Then our estimate for ψ implies that $\gamma\varphi \in C^{4,\alpha}_{\delta+2-2n}(\mathbf{C}^n \setminus B_1)$, and so

$$\varphi \in C^{4,\alpha}_{\delta+2-2n}(\mathrm{Bl}_0\mathbf{C}^n).$$

This decay is enough to show that the following integration by parts is possible:

$$0 = \int_{\mathrm{Bl}_0\mathbf{C}^n} \varphi\mathcal{D}^*\mathcal{D}\varphi\,\eta^n = \int_{\mathrm{Bl}_0\mathbf{C}^n} |\mathcal{D}\varphi|^2\,\eta^n,$$

so $\mathcal{D}\varphi = 0$. This implies that $\mathrm{grad}^{1,0}\varphi$ is a holomorphic vector field on $\mathrm{Bl}_0\mathbf{C}^n$, and it gives rise to a holomorphic vector field v on the complement $\mathbf{C}^n \setminus B$ of the unit ball B. The components of v are holomorphic functions, which by Hartog's theorem can be extended to all of $\mathbf{C}^n$. At the same time the components decay at infinity, so we must have $v = 0$. This implies that φ is constant, but φ also decays at infinity, so $\varphi = 0$. □

The analogous result for the operator $\tilde{L}_\omega$ on M_p is the following.

Proposition 8.10. *If $\delta > 4-2n$, then*

$$\tilde{L}_\omega : C^{4,\alpha}_\delta(M_p) \to C^{0,\alpha}_{\delta-4}(M_p)$$

has trivial kernel.

Proof. We can assume that δ is not an integer in order to avoid indicial roots in the argument below. Recall that we fixed a point q different from p and that

$$\tilde{L}_\omega(\varphi) = -\mathcal{D}^*\mathcal{D}\varphi - \varphi(q).$$

Suppose that $\tilde{L}_\omega(\varphi) = 0$ and $\varphi \in C^{4,\alpha}_\delta(M_p)$. We let γ be a cutoff function supported in B_1, equal to 1 in $B_{1/2}$ so that we can think of $\gamma\varphi$ as a function on $\mathbf{R}^{2n} \setminus \{0\}$. Again we compare $\mathcal{D}^*\mathcal{D}$ to the Euclidean Δ^2, and we obtain

$$\Delta^2(\gamma\varphi) \in C^{0,\alpha}_{\delta-2},$$

as long as $\delta < 2$. From Exercise 8.8 we have a $\psi \in C^{4,\alpha}_{\delta+2}(\mathbf{R}^{2n} \setminus \{0\})$ such that

$$\Delta^2(\gamma\varphi - \psi) = 0.$$

We have $\gamma\varphi - \psi \in C^{4,\alpha}_\delta(B_1 \setminus \{0\})$, and since there are no indicial roots in $(4-2n, 0)$, we obtain

$$\gamma\varphi - \psi \in C^{4,\alpha}_0(B_1 \setminus \{0\}).$$

Note that when $n = 2$ we already have better decay than this by assumption. From our decay estimate for ψ we obtain $\varphi \in C^{4,\alpha}_{\delta+2}(M_p)$ if $\delta + 2 < 0$ or $\varphi \in C^{4,\alpha}_0(M_p)$ otherwise. We can repeat the argument with $\delta + 2$ instead of δ if necessary, to eventually obtain $\varphi \in C^{4,\alpha}_0(M_p)$ in either case.

From this it follows that φ actually extends smoothly across p, and we have already seen that there are no smooth functions on M in the kernel of $\tilde{L}_\omega$. □

Remark 8.11. Although we will not use this fact, it is interesting to note that by Theorem 8.6 the previous two propositions imply that

$$\tilde{L}_\omega : C^{4,\alpha}_\delta(M_p) \to C^{0,\alpha}_{\delta-4}(M_p)$$

is surjective for $\delta < 0$, while

$$L_\eta : C^{4,\alpha}_\delta(\mathrm{Bl}_0\mathbf{C}^n) \to C^{0,\alpha}_{\delta-4}(\mathrm{Bl}_0\mathbf{C}^n)$$

is surjective for $\delta > 4-2n$. In particular both maps are isomorphisms for $\delta \in (4-2n, 0)$.

8.2.3. Weighted spaces on $\mathrm{Bl}_p M$**.** The weighted spaces that we use on $\mathrm{Bl}_p M$ are essentially glued versions of the spaces defined on M_p and on $\mathrm{Bl}_0\mathbf{C}^n$. Recall that we have chosen normal coordinates z^i around p, defined for $|z| < 1$. To obtain $\mathrm{Bl}_p M$ we are gluing in a scaled down version of $\mathrm{Bl}_0\mathbf{C}^n$. In terms of the coordinates w^i on $\mathrm{Bl}_0\mathbf{C}^n$ we perform the gluing by identifying the annuli

$$\{r_\varepsilon < |z| < 2r_\varepsilon\} = \{\varepsilon^{-1}r_\varepsilon < |w| < 2\varepsilon^{-1}r_\varepsilon\},$$
$$z = \varepsilon w.$$

There are three regions of the manifold $\mathrm{Bl}_p M$:

- $M \setminus B_1$: Here the coordinates z^i are not defined, but we can think of it as the region where $|z| \geqslant 1$.
- $B_1 \setminus B_\varepsilon$: This region can either be thought of as a subset of M, where $\varepsilon \leqslant |z| < 1$, or also as a subset of $\mathrm{Bl}_0\mathbf{C}^n$, where $1 \leqslant |w| < \varepsilon^{-1}$.
- B_ε: Here the coordinates z^i are again not defined, and this region should be thought of as the subset of $\mathrm{Bl}_0\mathbf{C}^n$, where $|w| < 1$. Note that this region is not actually a ball since it contains the exceptional divisor.

We define the weighted Hölder norms as follows. Suppose that $f : \mathrm{Bl}_p M \to \mathbf{R}$ and fix a weight δ. For $r \in (\varepsilon, 1/2)$ define

$$f_r : B_2 \setminus B_1 \to \mathbf{R}$$
$$f_r(z) = r^{-\delta} f(rz)$$

and also let

$$f_\varepsilon : \widetilde{B}_1 \subset \mathrm{Bl}_0\mathbf{C}^n \to \mathbf{R}$$
$$f_\varepsilon(w) = \varepsilon^{-\delta} f(\varepsilon w),$$

where $\widetilde{B}_1$ is the subset of $\mathrm{Bl}_0\mathbf{C}^n$ where $|w| < 1$. We are abusing notation, writing $w \mapsto \varepsilon w$ for the map identifying $\widetilde{B}_1$ with $B_\varepsilon \subset \mathrm{Bl}_p M$. The weighted norm is defined as

$$\|f\|_{C^{k,\alpha}_\delta(\mathrm{Bl}_p M)} = \|f\|_{C^{k,\alpha}(M\setminus B_1)} + \sup_{\varepsilon<r<1/2} \|f_r\|_{C^{k,\alpha}(B_2\setminus B_1)} + \|f_\varepsilon\|_{C^{k,\alpha}(\widetilde{B}_1)}.$$

On $M \setminus B_1$ and $\widetilde{B}_1$ we are measuring the Hölder norms with respect to fixed background metrics (or alternatively with respect to fixed coverings by charts). On $B_2 \setminus B_1$ we use the standard Euclidean metric.

Recall the cutoff functions γ_i that we used in Subsection 8.1.3. Since $\nabla\gamma_i$ is supported on $B_{2r_\varepsilon} \setminus B_{r_\varepsilon}$ and since it is of order r_ε^{-1} on this annulus, one can check that

$$\|\gamma_i\|_{C^{4,\alpha}_0(\mathrm{Bl}_p M)} \leqslant c$$

for some constant c independent of ε. One use for these cutoff functions is that if f is a function on $\mathrm{Bl}_p M$, then $\gamma_1 f$ and $\gamma_2 f$ can be naturally thought of as functions on M_p and $\mathrm{Bl}_0\mathbf{C}^n$, respectively. Using these, an equivalent weighted norm could be defined as

$$\|f\|_{C^{k,\alpha}_\delta(\mathrm{Bl}_p M)} = \|\gamma_1 f\|_{C^{k,\alpha}_\delta(M_p)} + \varepsilon^{-\delta}\|\gamma_2 f\|_{C^{k,\alpha}_\delta(\mathrm{Bl}_0\mathbf{C}^n)}. \tag{8.6}$$

Note that the spaces $C^{k,\alpha}_\delta$ themselves do not depend on δ, as they all consist of functions on $\mathrm{Bl}_p M$ which are locally in $C^{k,\alpha}$. The weight δ only affects the norm. There are simple inequalities relating the norms for different weights:

$$\|f\|_{C^{k,\alpha}_{\delta'}} \leqslant \begin{cases} \|f\|_{C^{k,\alpha}_\delta} & \text{if } \delta' \leqslant \delta, \\ \varepsilon^{\delta-\delta'}\|f\|_{C^{k,\alpha}_\delta} & \text{if } \delta' > \delta. \end{cases} \tag{8.7}$$

Lemma 8.12. *Let us write g_ε for the metric defined by ω_ε. We have the estimates*

$$\|g_\varepsilon\|_{C^{2,\alpha}_0(\mathrm{Bl}_p M)}, \|g_\varepsilon^{-1}\|_{C^{2,\alpha}_0(\mathrm{Bl}_p M)} \leqslant C,$$

for the components of the metric g_ε and its inverse, where C is independent of ε.

Proof. We are measuring the Hölder norms of the components of g_ε and g_ε^{-1}. We can deal with the three regions of $\mathrm{Bl}_p M$ separately. For instance, to deal with the annulus $B_{2r}\setminus B_r$, we need to pull back the components of g_ε to $B_2\setminus B_1$ (note that this is different from pulling back the metric itself, which would introduce a factor of r^2). The result follows since by the construction in Subsection 8.1.3 these pulled back metrics are uniformly equivalent to the Euclidean metric. □

Lemma 8.13. *There are constants $c_0, C_1 > 0$ with the following property. If $\|\varphi\|_{C^{4,\alpha}_2} < c_0$, then $\omega_\varphi = \omega_\varepsilon + \sqrt{-1}\partial\bar\partial\varphi$ is positive, and the corresponding metric g_φ satisfies*

$$\|g_\varphi - g_\varepsilon\|_{C^{2,\alpha}_{\delta-2}}, \|g_\varphi^{-1} - g_\varepsilon^{-1}\|_{C^{2,\alpha}_{\delta-2}}, \|\mathrm{Rm}_{g_\varphi} - \mathrm{Rm}_{g_\varepsilon}\|_{C^{0,\alpha}_{\delta-4}} < C_1\|\varphi\|_{C^{4,\alpha}_\delta},$$
$$\|L_{\omega_\varphi} - L_{\omega_\varepsilon}\|_{C^{4,\alpha}_\delta \to C^{0,\alpha}_{\delta-4}} < C_1\|\varphi\|_{C^{4,\alpha}_2},$$

where in the second line we are measuring the operator norm.

Proof. We have $\|\sqrt{-1}\partial\bar\partial\varphi\|_{C^{2,\alpha}_0} < C_2\|\varphi\|_{C^{4,\alpha}_2}$, so as long as c_0 is small enough, the form ω_φ is positive, and moreover we have

$$\|g_\varphi^{-1}\|_{C^{2,\alpha}_0} < 2C,$$

where C is as in the previous lemma. The required estimates can then be obtained by straightforward calculations using multiplication properties of the weighted norms. For example

$$g_\varphi^{-1} - g_\varepsilon^{-1} = g_\varphi^{-1}(g_\varepsilon - g_\varphi)g_\varepsilon^{-1},$$

and so

$$\|g_\varphi^{-1} - g_\varepsilon^{-1}\|_{C^{2,\alpha}_{\delta-2}} \leqslant \|g_\varphi^{-1}\|_{C^{2,\alpha}_0}\|\sqrt{-1}\partial\bar\partial\varphi\|_{C^{2,\alpha}_{\delta-2}}\|g_\varepsilon^{-1}\|_{C^{2,\alpha}_0} < C_3\|\varphi\|_{C^{4,\alpha}_\delta}. \qquad \square$$

The heart of the matter is the following result, which gives good bounds on the inverse of the linearization of our equation. The proof uses a contradiction argument together with rescaling.

Theorem 8.14. *For $\delta \in (-1, 0)$ there is a constant K such that for sufficiently small ε the operator*

$$\begin{aligned} F_\varepsilon : C^{4,\alpha}_\delta(\mathrm{Bl}_pM) &\to C^{0,\alpha}_{\delta-4}(\mathrm{Bl}_pM) \\ \varphi &\mapsto \mathcal{D}^*_{\omega_\varepsilon}\mathcal{D}_{\omega_\varepsilon}\varphi - \varphi(q) \end{aligned}$$

is invertible and we have a bound $\|F_\varepsilon^{-1}\| < K\varepsilon^\delta$ for its inverse. When the dimension $n > 2$, then we have a bound $\|F_\varepsilon^{-1}\| < K$ independent of ε.

Proof. We will only focus on the $n = 2$ case since when $n > 2$ the proof is easier. Note that we already know that F_ε is invertible from our assumption that M has no holomorphic vector fields. This implies that Bl_pM also has no holomorphic vector fields (see Exercise 8.1), and so the kernel of $\mathcal{D}^*_{\omega_\varepsilon}\mathcal{D}_{\omega_\varepsilon}$ consists of the constants. It follows from this that $\ker F_\varepsilon$ is trivial, and since it has index zero, it is an isomorphism. What we need to show is that the inverse of F_ε has the required bounds.

The Schauder estimates imply that there exists a constant C independent of ε such that

$$\|\varphi\|_{C^{4,\alpha}_\delta} \leqslant C(\|\varphi\|_{C^0_\delta} + \|F_\varepsilon(\varphi)\|_{C^{0,\alpha}_{\delta-4}}). \tag{8.8}$$

This can be seen as follows. Let us define $\rho : \mathrm{Bl}_pM \to \mathbf{R}$ by

$$\rho(z) = \begin{cases} 1 & \text{if } x \in M \setminus B_1, \\ |z| & \text{if } x \in B_1 \setminus B_\varepsilon, \\ \varepsilon & \text{if } x \in B_\varepsilon, \end{cases} \tag{8.9}$$

where we are identifying $\mathrm{Bl}_pM \setminus B_\varepsilon$ with $M \setminus B_\varepsilon$ as before to make sense of the distance $|z|$. The key point is that we can choose a small scale $r > 0$, independently of ε, such that for each point $z \in \mathrm{Bl}_pM$ the ball of radius $r\rho(z)$ around z with respect to ω_ε, when scaled to unit size, is close to the Euclidean unit ball. Here "close" means that the components of the metrics can be made as close as we wish in any C^k-norm by choosing r sufficiently small. Note that for this to be true it is important that the Burns-Simanca metric η is asympotically flat, but otherwise it follows directly from the construction of ω_ε. Estimate (8.8) now follows by applying the usual Schauder estimates to these balls of radii $r\rho(z)$ scaled to unit size.

Our goal is to show that inequality (8.8) also holds without the $\|\varphi\|_{C^0_\delta}$ term up to replacing C by $K\varepsilon^\delta$. We argue by contradiction. If the estimate

$$\|\varphi\|_{C^{4,\alpha}_\delta} \leqslant K\varepsilon^\delta \|F_\varepsilon(\varphi)\|_{C^{0,\alpha}_{\delta-4}}$$

does not hold for sufficiently small ε for any K, then there is a sequence $\varepsilon_i \to 0$ and corresponding functions φ_i such that

$$\|\varphi_i\|_{C^{4,\alpha}_\delta} = 1, \text{ but } \|F_{\varepsilon_i}(\varphi_i)\|_{C^{0,\alpha}_{\delta-4}} < \frac{1}{i}\varepsilon_i^{-\delta}. \tag{8.10}$$

Note that each φ_i is a function on $\mathrm{Bl}_p M$, but the weighted norms depend on i. We will obtain a contradiction by extracting limits of the φ_i on three different regions: on M_p, $\mathrm{Bl}_p M$, and the "neck region" $\mathbf{C}^2 \setminus \{0\}$.

First consider the integral of $F_{\varepsilon_i}(\varphi_i)$ over $\mathrm{Bl}_p M$. From (8.10) we can estimate

$$\left| \int_{\mathrm{Bl}_p M} F_{\varepsilon_i}(\varphi_i)\, \omega_{\varepsilon_i}^2 \right| < C\frac{1}{i},$$

but the integral of $\mathcal{D}^*\mathcal{D}\varphi$ vanishes, so this implies

$$|\varphi_i(q)| < C\frac{1}{i}, \tag{8.11}$$

for some constant C.

We now consider the limit on M_p. We can think of the φ_i as functions on $M \setminus B_{\varepsilon_i}$, i.e. on larger and larger subsets of M_p. The uniform $C^{4,\alpha}_\delta$-bounds imply that up to choosing a subsequence, we can assume that $\varphi_i \to \varphi_\infty$ locally in $C^{4,\alpha'}_\delta$ for some $\alpha' < \alpha$, and since $\varphi_i(q) \to 0$, the function φ_∞ on M_p satisfies $\varphi_\infty(q) = 0$ and

$$\mathcal{D}^*_\omega \mathcal{D}_\omega \varphi_\infty = 0. \tag{8.12}$$

At this point, if we were working in dimensions $n > 2$, then we would obtain $\varphi_\infty = 0$ from Proposition 8.10. When $n = 2$ we need to work a bit harder. By an argument similar to that in the proof of Proposition 8.10 we obtain that

$$\varphi_\infty = \psi + a \log|z|, \tag{8.13}$$

where $\psi \in C^{4,\alpha'}_0$ and $a \in \mathbf{R}$. The point is that the only biharmonic functions on $B_1 \setminus \{0\}$ with growth rate in $[\delta, 0]$ are $\log|z|$ and the constants. We want to show that $a = 0$.

One can see this by integrating equation (8.12) over $M \setminus B_r$ for sufficiently small r and integrating by parts. The conceptual reason is that we can think of φ_∞ as giving a solution of the equation

$$\mathcal{D}_\omega^* \mathcal{D}_\omega \varphi_\infty = ac\delta_p,$$

on all of M in the sense of distributions, where δ_p denotes the Delta function at p and c is a non-zero constant. But pairing both sides of the equation with the constant 1, we obtain a contradiction unless $a = 0$. It follows therefore from (8.13) that $\varphi_\infty \in C_0^{4,\alpha'}$. As in Proposition 8.10 it follow then that φ_∞ extends as a smooth function on M, so we obtain $\varphi_\infty = 0$.

From (8.10) and (8.8) we find that $\|\varphi_i\|_{C_\delta^0}$ is bounded above and is bounded away from zero. By scaling the functions φ_i by suitable factors, we can obtain new functions ψ_i satisfying the following:

$$\begin{aligned} &\|\psi_i\|_{C_\delta^0} = 1, \quad \|\psi_i\|_{C_\delta^{4,\alpha}} < C, \\ &\|\mathcal{D}_{\omega_{\varepsilon_i}}^* \mathcal{D}_{\omega_{\varepsilon_i}} \psi_i\|_{C_{\delta-4}^{0,\alpha}} \to 0, \quad \text{and } \psi_i \to 0 \text{ locally on } M_p \text{ in } C^{4,\alpha}. \end{aligned} \tag{8.14}$$

We now need to examine the point $q_i \in \mathrm{Bl}_p M$, where the function $\rho_i^{-\delta}\psi_i$ achieves its maximum, i.e.

$$\rho_i^{-\delta}(q_i)\psi_i(q_i) = 1,$$

with ρ_i denoting the function in (8.9) corresponding to ε_i. Since $\psi_i \to 0$ locally on M_p, we must have $\rho_i(q_i) \to 0$, and there are two different cases depending on whether $\varepsilon_i^{-1}\rho_i(q_i)$ is bounded or not.

If $\varepsilon_i^{-1}\rho_i(q_i) < R$ for some R and for all i, then under the identification of $B_1 \subset \mathrm{Bl}_p M$ with the ball $B_{\varepsilon^{-1}} \subset \mathrm{Bl}_0 \mathbf{C}^2$ (see Subsection 8.2.3), the points q_i are inside $B_R \subset \mathrm{Bl}_0 \mathbf{C}^2$. Up to choosing a subsequence, we can assume that $q_i \to z_\infty \in B_R$. Moreover, using (8.6) we can think of $\varepsilon_i^{-\delta}\psi_i$ as functions on larger and larger subsets of $\mathrm{Bl}_0\mathbf{C}^2$, bounded in $C_\delta^{4,\alpha}$, so up to choosing a further subsequence, we can assume that $\varepsilon_i^{-\delta}\psi_i \to \psi_\infty$ locally in $C^{4,\alpha'}$ on $\mathrm{Bl}_0\mathbf{C}^2$ and that $\psi_\infty \in C_\delta^{4,\alpha'}$. From (8.14) we obtain

$$\begin{aligned} \psi_\infty(z_\infty) &\geqslant R^\delta, \\ \mathcal{D}_\eta^* \mathcal{D}_\eta \psi_\infty &= 0. \end{aligned}$$

For the second equation note that $\varepsilon_i^{-2}\omega_{\varepsilon_i}$ converges to the LeBrun-Simanca metric η under our identifications. This contradicts the fact that there are no non-zero elements in the kernel of $\mathcal{D}_\eta^*\mathcal{D}_\eta$ which decay at infinity, by Proposition 8.9.

The last case to examine is when $\varepsilon_i^{-1}\rho_i(q_i)$ is unbounded but $\rho_i(q_i) \to 0$. In particular in this case we have $\rho_i(q_i) = |q_i|$ from the definition of ρ_i. We can choose sequences $r_i \to 0$ and $R_i \to \infty$ such that the annulus

$$A_i = B_{R_i|q_i|} \setminus B_{r_i|q_i|}$$

is contained in smaller and smaller balls around p when thought of as a region in M_p, while it is also in the complement of larger and larger balls in $\mathrm{Bl}_0\mathbf{C}^2$ when thought of as a set there. More precisely R_i, r_i satisfy

$$R_i|q_i| \to 0, \quad r_i|q_i|\varepsilon_i^{-1} \to \infty.$$

We identify the annulus A_i with the annulus $B_{R_i} \setminus B_{r_i} \subset \mathbf{C}^2 \setminus \{0\}$ by scaling, with corresponding metric $|q_i|^{-2}\omega_{\varepsilon_i}$. These scaled metrics converge to the flat metric on $\mathbf{C}^2 \setminus \{0\}$ locally uniformly in any C^k-norm. The points q_i correspond to points on the unit sphere by this identification, so up to choosing a subsequence we can assume $q_i \to q_\infty$. We can think of the functions $|q_i|^{-\delta}\psi_i$ as functions on larger and larger subsets of $\mathbf{C}^2 \setminus \{0\}$, with a uniform $C^{4,\alpha}_\delta$-bound, so choosing a further subsequence, we have $|q_i|^{-\delta}\psi_i \to \psi_\infty$ locally in $C^{4,\alpha'}$. The limit satisfies $\psi_\infty \in C^{4,\alpha'}_\delta$ and

$$\psi_\infty(q_\infty) = 1,$$
$$\Delta^2\psi_\infty = 0.$$

We used that $\mathcal{D}^*\mathcal{D} = \Delta^2$ for the flat metric. This contradicts Theorem 8.3, and so the proof is complete. □

The operator that we really want to invert is $\tilde{L}_\varepsilon$ rather than F_ε. The difference between the two is the operator

$$T : \varphi \mapsto g_\varepsilon^{j\bar{k}}\nabla_j S(\omega_\varepsilon)\nabla_{\bar{k}}\varphi.$$

From the proof of Lemma 8.19 below we obtain the estimate

$$\|\nabla S(\omega_\varepsilon)\|_{C^{0,\alpha}_{-3}} < Cr_\varepsilon^2$$

for some constant C, which means that we have a bound

$$\|T\|_{C^{4,\alpha}_\delta \to C^{0,\alpha}_{\delta-4}} < Cr_\varepsilon^2$$

on the operator norm of T with a larger C. This implies that from the bound for the inverse of F_ε in Theorem 8.14 we obtain a similar bound for the inverse of $\tilde{L}_\varepsilon$ for sufficiently small ε.

Exercise 8.15. Is the dependence on ε in the bound $\|F_\varepsilon^{-1}\| < K\varepsilon^\delta$ in Theorem 8.14 sharp in the case when $n = 2$?

Exercise 8.16. Prove an analogous result to Theorem 8.14 for $\delta \in (0, 1)$. Do you obtain a bound on the inverse independent of ε? What about when $\delta = 0$?

Remark 8.17. There are several alternative approaches to the proof of a result such as Theorem 8.14, and in different geometric problems one method may have advantages over others. The approach above was used by

Biquard-Rollin [**16**] for the similar problem of smoothing out singular cscK metrics.

Another possibility is to use the inverses of the linear operators in Remark 8.11 together with suitable cutoff functions to construct an approximate inverse of $\tilde{L}_\varepsilon$ on Bl_pM. For sufficiently small ε this can be perturbed to a genuine inverse. This is the approach in [**108**] for instance.

Yet another approach is to work on manifolds with boundary, $M \setminus B_{r_\varepsilon}$ and $B_{\varepsilon^{-1}r_\varepsilon} \subset \mathrm{Bl}_0\mathbf{C}^n$, and glue inverses of the linear operators by matching boundary data. This method can be applied directly to the non-linear operator, as was done by Arezzo-Pacard [**3**].

8.3. Solving the non-linear equation when $n > 2$

We are now ready to solve the equation

$$S(\omega_\varepsilon + \sqrt{-1}\partial\bar{\partial}\varphi) - S(\omega) - \varphi(q) = 0, \tag{8.15}$$

for sufficiently small ε where q is a point outside the unit ball around p. We will choose $\delta < 0$ to be very close to 0. Writing

$$S(\omega_\varepsilon + \sqrt{-1}\partial\bar{\partial}\varphi) = S(\omega_\varepsilon) + L_{\omega_\varepsilon}(\varphi) + Q_{\omega_\varepsilon}(\varphi), \tag{8.16}$$

equation (8.15) is equivalent to

$$S(\omega_\varepsilon) - S(\omega) + L_{\omega_\varepsilon}(\varphi) - \varphi(q) + Q_{\omega_\varepsilon}(\varphi) = 0.$$

From Theorem 8.14 we know that the operator

$$\tilde{L}_{\omega_\varepsilon}(\varphi) := L_{\omega_\varepsilon}(\varphi) - \varphi(q)$$

has an inverse, in terms of which our equation can be rewritten as

$$\varphi = \tilde{L}_{\omega_\varepsilon}^{-1}\Big(S(\omega) - S(\omega_\varepsilon) - Q_{\omega_\varepsilon}(\varphi)\Big).$$

Let us define the operator $\mathcal{N}$ by

$$\begin{aligned}\mathcal{N} : C^{4,\alpha}_\delta(\mathrm{Bl}_pM) &\to C^{4,\alpha}_\delta(\mathrm{Bl}_pM)\\ \varphi &\mapsto \tilde{L}_{\omega_\varepsilon}^{-1}\Big(S(\omega) - S(\omega_\varepsilon) - Q_{\omega_\varepsilon}(\varphi)\Big).\end{aligned}$$

Equation (8.15) is then equivalent to the fixed point problem $\varphi = \mathcal{N}(\varphi)$.

The following lemma shows that $\mathcal{N}$ is a contraction on a suitable set.

Lemma 8.18. *When $n > 2$, there is a constant $c_1 > 0$ such that if*

$$\|\varphi\|_{C^{4,\alpha}_2}, \|\psi\|_{C^{4,\alpha}_2} \leqslant c_1,$$

then

$$\|\mathcal{N}(\varphi) - \mathcal{N}(\psi)\|_{C^{4,\alpha}_\delta} \leqslant \frac{1}{2}\|\varphi - \psi\|_{C^{4,\alpha}_\delta}.$$

Proof. We have

$$\mathcal{N}(\varphi) - \mathcal{N}(\psi) = \tilde{L}_{\omega_\varepsilon}^{-1}(Q_{\omega_\varepsilon}(\psi) - Q_{\omega_\varepsilon}(\varphi)).$$

By the mean value theorem there is a $t \in [0,1]$ such that $\chi = t\varphi + (1-t)\psi$ satisfies

$$Q_{\omega_\varepsilon}(\psi) - Q_{\omega_\varepsilon}(\varphi) = DQ_{\omega_\varepsilon,\chi}(\psi - \varphi).$$

Differentiating (8.16) at χ we have

$$DQ_{\omega_\varepsilon,\chi} = L_{\omega_\chi} - L_{\omega_\varepsilon},$$

so from Lemma 8.13 we know that if $\|\chi\|_{C_2^{4,\alpha}} < c_0$, then

$$\begin{aligned} \|Q_{\omega_\varepsilon}(\psi) - Q_{\omega_\varepsilon}(\varphi)\|_{C_{\delta-4}^{0,\alpha}} &< C\|\chi\|_{C_2^{4,\alpha}}\|\psi - \varphi\|_{C_\delta^{4,\alpha}} \\ &\leq C\{\|\varphi\|_{C_2^{4,\alpha}} + \|\psi\|_{C_2^{4,\alpha}}\}\|\psi - \varphi\|_{C_\delta^{4,\alpha}}. \end{aligned} \tag{8.17}$$

Since $\tilde{L}_{\omega_\varepsilon}^{-1}$ is bounded by Theorem 8.14, the result follows once c_1 is chosen small enough. □

We also need to know how good our approximate solution ω_ε is.

Lemma 8.19. *For sufficiently small ε we have*

$$\|S(\omega_\varepsilon) - S(\omega)\|_{C_{\delta-4}^{0,\alpha}} \leqslant Cr_\varepsilon^{4-\delta}$$

for some constant C.

Proof. We examine three different regions of Bl_pM. On $M\backslash B_{2r_\varepsilon}$ the metrics ω_ε and ω are equal, so $S(\omega_\varepsilon) - S(\omega) = 0$.

On $B_{2r_\varepsilon} \setminus B_{r_\varepsilon}$, in terms of the Euclidean metric ω_E we have

$$\omega_E = \omega_\varepsilon + \sqrt{-1}\partial\bar{\partial}\varphi,$$

where $\varphi = O(|z|^4)$. It follows that

$$\|\varphi\|_{C_\delta^{4,\alpha}(B_{2r_\varepsilon}\setminus B_{r_\varepsilon})} \leqslant C_1 r_\varepsilon^{4-\delta}$$

for some constant C_1. Using this for $\delta = 2$ as well, Lemma 8.13 implies that

$$\|S(\omega_\varepsilon) - 0\|_{C_{\delta-4}^{0,\alpha}(B_{2r_\varepsilon}\setminus B_{r_\varepsilon})} \leqslant C_2 r_\varepsilon^{4-\delta},$$

for sufficiently small ε. Since $S(\omega)$ is a fixed constant, we also have

$$\|S(\omega)\|_{C_{\delta-4}^{0,\alpha}(B_{2r_\varepsilon}\setminus B_{r_\varepsilon})} \leqslant C_3 r_\varepsilon^{4-\delta},$$

so this takes care of the annulus $B_{2r_\varepsilon} \setminus B_{r_\varepsilon}$.

On B_{r_ε} we have $S(\omega_\varepsilon) = 0$, and again

$$\|S(\omega)\|_{C_{\delta-4}^{0,\alpha}(B_{r_\varepsilon})} \leqslant C_3 r_\varepsilon^{4-\delta}.$$ □

We can finally put the pieces together.

Proposition 8.20. *Assume $n > 2$. Using the constant c_1 from Lemma 8.18, let*

$$\mathcal{U} = \left\{ \varphi \in C^{4,\alpha}_\delta \,:\, \|\varphi\|_{C^{4,\alpha}_\delta} \leqslant c_1 \varepsilon^{2-\delta} \right\} \subset C^{4,\alpha}_\delta(\mathrm{Bl}_p M).$$

If ε is sufficiently small, then $\mathcal{N}$ is a contraction on $\mathcal{U}$, and $\mathcal{N}(\mathcal{U}) \subset \mathcal{U}$. In particular $\mathcal{N}$ has a fixed point, which gives a cscK metric on $\mathrm{Bl}_p M$ in the Kähler class $\pi^*[\omega] - \varepsilon^2[E]$.

Proof. From the comparison in (8.7) between the weighted norms, we have $\|\varphi\|_{C^{4,\alpha}_2} \leqslant c_1$ if $\varphi \in \mathcal{U}$. From Lemma 8.18 it follows then that $\mathcal{N}$ is a contraction on $\mathcal{U}$ and in addition

$$\begin{aligned} \|\mathcal{N}(\varphi)\|_{C^{4,\alpha}_\delta} &\leqslant \|\mathcal{N}(\varphi) - \mathcal{N}(0)\|_{C^{4,\alpha}_\delta} + \|\mathcal{N}(0)\|_{C^{4,\alpha}_\delta} \\ &\leqslant \frac{1}{2}\|\varphi\|_{C^{4,\alpha}_\delta} + \|\mathcal{N}(0)\|_{C^{4,\alpha}_\delta}. \end{aligned} \tag{8.18}$$

From Lemma 8.19 we have

$$\|\mathcal{N}(0)\|_{C^{4,\alpha}_\delta} \leqslant C\|S(\omega_\varepsilon) - S(\omega)\|_{C^{0,\alpha}_{\delta-4}} \leqslant C' r_\varepsilon^{4-\delta}.$$

From the definition of r_ε,

$$r_\varepsilon^{4-\delta} = \varepsilon^{(4-\delta)\frac{n-1}{n}},$$

and if δ is close to 0 and $n > 2$, then

$$(4-\delta)\frac{n-1}{n} > 2 - \delta. \tag{8.19}$$

It follows that if ε is sufficiently small, we have

$$\|\mathcal{N}(0)\|_{C^{4,\alpha}_\delta} \leqslant \frac{1}{2} c_1 \varepsilon^{2-\delta}.$$

From (8.18) we then have $\mathcal{N}(\varphi) \in \mathcal{U}$. This completes the proof. □

8.4. The case when $n = 2$

We have given the proof of Theorem 8.2 in the case when the dimension $n > 2$. The most apparent reason for this restriction is that the best range of weights for the linear analysis is $\delta \in (4-2n, 0)$, and when $n = 2$ this set is empty. We have seen in Theorem 8.14 that when $n = 2$ we can still work with $\delta \in (-1, 0)$, but we lose the uniform control of the inverse operator. Another issue comes from inequality (8.19). One can check that even by choosing $r_\varepsilon = \varepsilon^\alpha$ for different α, the corresponding inequality cannot be satisfied when $n = 2$. For this, note that the estimate in Lemma 8.19 becomes worse when we do not have $\alpha = \frac{n-1}{n}$. One way to overcome these problems is to construct a better approximate solution than the one constructed in Subsection 8.1.3.

First of all, as we mentioned in Subsection 8.1.2, the Burns-Simanca metric has the form

$$\eta = \sqrt{-1}\partial\overline{\partial}(|w|^2 + \log|w|^2)$$

on $\mathrm{Bl}_0\mathbf{C}^2$. We can follow the construction in Subsection 8.1.3 to obtain the metric ω_ε. On the annulus $B_{2r_\varepsilon} \setminus B_{r_\varepsilon}$ the metric is given by

$$\sqrt{-1}\partial\overline{\partial}\Big(|z|^2 + \gamma_1(z)\varphi_1(z) + \varepsilon^2\gamma_2(z)\log|\varepsilon^{-1}z|\Big).$$

We will construct a new metric

$$\widetilde{\omega}_\varepsilon = \omega_\varepsilon + \varepsilon^2\sqrt{-1}\partial\overline{\partial}\Big(\gamma_1(z)(\Gamma(z) - \log\varepsilon)\Big), \tag{8.20}$$

where the leading term in $\Gamma(z)$ is $\log|z|$ and $\Gamma(z)$ solves the equation

$$\mathcal{D}^*_\omega\mathcal{D}_\omega\Gamma = c \tag{8.21}$$

on M_p, where c is a constant.

To show that such a Γ exists, we again use that $\mathcal{D}^*_\omega\mathcal{D}_\omega$ is a small perturbation of the Euclidean Δ^2_{Euc}. Using the notation from Subsection 8.1.3, in normal coordinates around p the metric ω is given by

$$\omega = \sqrt{-1}\partial\overline{\partial}\Big(|z|^2 + \varphi_1(z)\Big),$$

where $\varphi_1 = O(|z|^4)$. For small $r > 0$, let γ be a cutoff function supported in B_{2r}, equal to 1 on B_r, and define the metric

$$\tilde{\omega} = \sqrt{-1}\partial\overline{\partial}\Big(|z|^2 + \gamma(z)\varphi_1(z)\Big)$$

on $\mathbf{R}^{2n} \setminus \{0\}$, which is flat outside B_{2r}. For any $\delta \in \mathbf{R}$ we have a constant C such that

$$\|(\Delta^2_{\mathrm{Euc}} - \mathcal{D}^*_{\tilde{\omega}}\mathcal{D}_{\tilde{\omega}})f\|_{C^{0,\alpha}_{\delta-4}} \leqslant Cr^2\|f\|_{C^{4,\alpha}_\delta},$$

for any $f : \mathbf{R}^{2n} \setminus \{0\} \to \mathbf{R}$. It follows that for sufficiently small r (perhaps depending on δ) the operator

$$\mathcal{D}^*_{\tilde{\omega}}\mathcal{D}_{\tilde{\omega}} : C^{4,\alpha}_\delta(\mathbf{R}^{2n} \setminus \{0\}) \to C^{0,\alpha}_{\delta-4}(\mathbf{R}^{2n} \setminus \{0\}) \tag{8.22}$$

is an isomorphism whenever δ is not an indicial root of Δ^2. Note that near the origin $\mathcal{D}^*_{\tilde{\omega}}\mathcal{D}_{\tilde{\omega}} = \mathcal{D}^*_\omega\mathcal{D}_\omega$.

By comparing $\mathcal{D}^*_\omega\mathcal{D}_\omega$ with Δ^2_{Euc} we find that

$$\mathcal{D}^*_{\tilde{\omega}}\mathcal{D}_{\tilde{\omega}}\log|z| \in C^{0,\alpha}_{-2}(\mathbf{R}^{2n} \setminus \{0\}),$$

and so using the isomorphism (8.22) we can find a function $\psi_1 \in C^{4,\alpha}_{2-\tau}$ for any small $\tau > 0$ such that near the point p we have

$$\mathcal{D}^*_\omega\mathcal{D}_\omega(\gamma\log|z| - \gamma\psi_1) = 0.$$

We are thinking of $\gamma\log|z|$ and $\gamma\psi_1$ as functions on M_p. Now using the isomorphism (8.4) between the usual Hölder spaces we can find a function $\psi_2 \in C_0^{4,\alpha}$ such that

$$\mathcal{D}_\omega^*\mathcal{D}_\omega(\gamma\log|z| - \gamma\psi_1 - \psi_2) = \psi_2(q) \quad \text{on } M_p.$$

We can then set $\Gamma = \gamma(\log|z| - \psi_1) - \psi_2$ to obtain (8.21), and Γ will be asymptotic to $\log|z|$ near p.

We now return to equation (8.20). The advantage of $\widetilde{\omega}_\varepsilon$ over ω_ε is that on the annulus $B_{2r_\varepsilon}\setminus B_{r_\varepsilon}$ it is given by

$$\widetilde{\omega}_\varepsilon = \sqrt{-1}\partial\overline{\partial}\Big(|z|^2 + \gamma_1(z)\varphi_1(z) + \varepsilon^2\log|z| + \varepsilon^2\gamma_1(z)\psi(z)\Big),$$

so now the $\log|z|$ term no longer needs to be multiplied by a cutoff function. The operator $\widetilde{L}_{\widetilde{\omega}_\varepsilon}$ is a sufficiently small perturbation of $\widetilde{L}_{\omega_\varepsilon}$, so that we have a bound $\|\widetilde{L}_{\widetilde{\omega}_\varepsilon}^{-1}\| < K\varepsilon^\delta$ from Theorem 8.14. We can then follow the arguments of Section 8.3 using the metric $\widetilde{\omega}_\varepsilon$ instead of ω_ε, except that we will solve the equation

$$S(\omega_\varepsilon + \sqrt{-1}\partial\overline{\partial}\varphi) - S(\omega) + \varepsilon^2 c - \varphi(q) = 0, \tag{8.23}$$

with c from (8.21).

Because of our slightly worse bound on the inverse of $\widetilde{L}_{\widetilde{\omega}_\varepsilon}$, instead of Lemma 8.18, we obtain the following, with the same proof.

Lemma 8.21. *For $\delta \in (-1,0)$, there is a constant $c_1 > 0$ such that if $\|\varphi\|_{C_2^{4,\alpha}}, \|\psi\|_{C_2^{4,\alpha}} \leqslant c_1\varepsilon^{-\delta}$, then*

$$\|\mathcal{N}(\varphi) - \mathcal{N}(\psi)\|_{C_\delta^{4,\alpha}} \leqslant \frac{1}{2}\|\varphi - \psi\|_{C_\delta^{4,\alpha}}.$$

The other ingredient we need is an estimate on how good our approximate solution is. At this point we fix $r_\varepsilon = \varepsilon^\alpha$, with

$$\alpha < \frac{2}{3},$$

and we will work with $\delta \in (-1,0)$ very close to 0. We obtain the following result analogous to Lemma 8.19. The estimate looks similar, but the point is that we are now able to choose a wider range of α, whereas the earlier result only applied when $\alpha = \frac{n-1}{n}$.

Lemma 8.22. *For sufficiently small ε and for $\delta \in (-1,0)$, we have*

$$\|S(\widetilde{\omega}_\varepsilon) + \varepsilon^2 c - S(\omega)\|_{C_{\delta-4}^{0,\alpha}} \leqslant Cr_\varepsilon^{4-\delta},$$

for some constant C.

Proof. Inside B_{r_ε} our metric is still scalar flat, so there we have

$$\|S(\widetilde{\omega}_\varepsilon) + \varepsilon^2 c - S(\omega)\|_{C^{0,\alpha}_{\delta-4}(B_{r_\varepsilon})} \leqslant C r_\varepsilon^{4-\delta}.$$

The main advantage of $\widetilde{\omega}_\varepsilon$ over ω_ε is on the region $B_{2r_\varepsilon} \setminus B_{r_\varepsilon}$. Here we can write

$$\widetilde{\omega}_\varepsilon = \omega_E + \sqrt{-1}\partial\overline{\partial}\xi$$

in terms of the Euclidean metric, where

$$\xi = \varepsilon^2 \log|z| + \gamma_1(z)\Big(\varphi_1(z) + \varepsilon^2\psi(z)\Big).$$

If $\alpha < 2/3$, then this implies

$$\xi = \varepsilon^2 \log|z| + O(|z|^4).$$

We can compute the scalar curvature as

$$S(\widetilde{\omega}_\varepsilon) = S(\omega_E) + L_{\omega_E}(\xi) + Q_{\omega_E}(\xi)$$

and note that $L_{\omega_E} = -\Delta_0^2$. Using (8.17) to control the Q term, we can compute that

$$\|S(\widetilde{\omega}_\varepsilon) + \varepsilon^2 c - S(\omega)\|_{C^{0,\alpha}_{\delta-4}(B_{2r_\varepsilon}\setminus B_{r_\varepsilon})} \leqslant C r_\varepsilon^{4-\delta}.$$

The remaining region is $M \setminus B_{2r_\varepsilon}$, on which we have

$$\widetilde{\omega}_\varepsilon = \omega + \varepsilon^2\sqrt{-1}\partial\overline{\partial}\Gamma.$$

It is helpful to break this region up further into $M \setminus B_1$ and regions of the form $B_{2r} \setminus B_r$ with $r > 2r_\varepsilon$. On $M \setminus B_1$ we have

$$S(\widetilde{\omega}_\varepsilon) = S(\omega) - \varepsilon^2\mathcal{D}_\omega^*\mathcal{D}_\omega\Gamma + Q_\omega(\varepsilon^2\Gamma), \tag{8.24}$$

which implies that

$$S(\widetilde{\omega}_\varepsilon) = S(\omega) - \varepsilon^2 c + O(\varepsilon^4).$$

From this it follows that

$$\|S(\widetilde{\omega}_\varepsilon) + \varepsilon^2 c - S(\omega)\|_{C^{0,\alpha}_{\delta-4}(M\setminus B_1)} \leqslant C r_\varepsilon^{4-\delta}$$

if δ is close to 0.

Finally, on the region $B_{2r} \setminus B_r$ with $r > 2r_\varepsilon$, we can use equation (8.24) to get

$$S(\widetilde{\omega}_\varepsilon) = S(\omega) - \varepsilon^2 c + O(\varepsilon^4 r^{-6}).$$

Once again, it follows that

$$\|S(\widetilde{\omega}_\varepsilon) + \varepsilon^2 c - S(\omega)\|_{C^{0,\alpha}_{\delta-4}(B_{2r}\setminus B_r)} \leqslant C r_\varepsilon^{4-\delta},$$

as long as $\alpha < 2/3$. □

We can now follow the proof of Proposition 8.20 to solve equation (8.23). Because of our slightly worse bound on the inverse of the linearized operator, we need to set

$$\mathcal{U} = \left\{ \varphi \in C^{4,\alpha}_\delta \, : \, \|\varphi\|_{C^{4,\alpha}_\delta} \leqslant c_1 \varepsilon^{2-2\delta} \right\},$$

and we need to show that

$$\|\mathcal{N}(0)\|_{C^{4,\alpha}_\delta} \leqslant \frac{1}{2} c_1 \varepsilon^{2-2\delta} \tag{8.25}$$

for sufficiently small ε. We have

$$\|\mathcal{N}(0)\|_{C^{4,\alpha}_\delta} \leqslant C\varepsilon^\delta \|S(\widetilde{\omega}_\varepsilon) + \varepsilon^2 c - S(\omega)\|_{C^{0,\alpha}_{\delta-4}} \leqslant C' \varepsilon^\delta r_\varepsilon^{4-\delta}.$$

If we choose $\alpha > 1/2$ (while also $\alpha < 2/3$), then

$$\delta + (4-\delta)\alpha > 2 - 2\delta$$

for δ sufficiently close to 0, so we can obtain (8.25) for sufficiently small ε. It follows that we can solve the non-linear equation once ε is small enough.

8.5. The case when M admits holomorphic vector fields

We have so far only dealt with manifolds M which have no holomorphic vector fields. If M does admit holomorphic vector fields, then the basic difficulty is that Proposition 8.10 no longer holds. Indeed the kernel and cokernel of $\mathcal{D}^*_\omega \mathcal{D}_\omega$ can be identified with holomorphic vector fields on M which have holomorphy potentials. This issue manifests itself in being unable to show that our linearized operator on $\mathrm{Bl}_p M$ is invertible, or at least we cannot find good bounds on its inverse. This is not merely a technical difficulty since when M has holomorphic vector fields, then $\mathrm{Bl}_p M$ may not admit a cscK metric even if M does.

The simplest example to consider is $M = \mathbf{CP}^2$ since then $\mathrm{Bl}_p M$ admits an extremal metric with non-constant scalar curvature (see Exercise 4.32). It follows that $\mathrm{Bl}_p M$ cannot admit a cscK metric since the Futaki invariant does not vanish. Of course one may hope then that if M admits an extremal metric, then so does $\mathrm{Bl}_p M$ in certain Kähler classes, but even this is not true. Indeed Levine's example in Remark 4.20 is a blow-up of $\mathbf{CP}^1 \times \mathbf{CP}^1$ in four points which does not admit an extremal metric in any Kähler class, with the structure of its automorphism group being the obstruction. A more general obstruction is given by K-stability.

The way that K-stability enters is that any one-parameter subgroup

$$\rho : \mathbf{C}^* \hookrightarrow \mathrm{Aut}(M)$$

gives rise to a test-configuration for Bl_pM no matter what ample line bundle we use on Bl_pM. The central fiber of this test-configuration is Bl_qM with

$$q = \lim_{t\to 0} \rho(t)\cdot p.$$

In view of Theorem 8.2 it is natural to expect that the only obstruction to the existence of an extremal metric on Bl_pM comes from such test-configurations, at least for Kähler classes that make the exceptional divisor sufficiently small. This expectation has been verified in the cscK case for $n > 2$ in [**103**]. At the time of this writing, the problem is still open for $n = 2$ and for general extremal metrics in any dimension.

In the rest of this section we will outline some of the modifications that need to be made to the analysis in the previous sections, although we will ignore certain technicalities to highlight the ideas that are applicable to other similar problems in geometric analysis. As we have seen, there are two main ingredients in the gluing method. We need to construct good approximate solutions, and we need to understand the linearized operator. For simplicity suppose that the dimension $n > 3$.

For constructing the approximate solutions, recall that we are trying to glue the extremal metric ω on M to the scaled Burns-Simanca metric $\varepsilon^2\eta$ along an annulus $B_{2r_\varepsilon}\setminus B_{r_\varepsilon}$ around $p\in M$. For better approximate solutions we need more precise expansions of these metrics than what we used before. In normal coordinates around p we have

$$\omega = \sqrt{-1}\partial\bar\partial\Big(|z|^2 + A_4(z) + A_5(z) + O(|z|^6)\Big),$$

where A_4 and A_5 are a quartic and quintic expression in z, respectively. One can also show (see Gauduchon [**57**]) that we have an expansion

$$\eta = \sqrt{-1}\partial\bar\partial\Big(|w|^2 - |w|^{4-2n} + d_1|w|^{2-2n} + d_2|w|^{6-4n} + O(|w|^{4-4n})\Big),$$

where d_1, d_2 are constants, so we are trying to glue ω to

$$\varepsilon^2\eta = \sqrt{-1}\partial\bar\partial\Big(|z|^2 - \varepsilon^{2n-2}|z|^{4-2n} + d_1\varepsilon^{2n}|z|^{2-2n} + d_2\varepsilon^{4n-4}|z|^{6-4n} + \cdots\Big),$$

with a change of variable $z = \varepsilon w$.

In constructing our crudest approximate solution ω_ε in Subsection 8.1.3, we multiplied every term apart from the $|z|^2$ with cutoff functions. To obtain better approximate solutions we would like to avoid cutting off certain other higher-order terms. For instance when obtaining the metric $\widetilde{\omega}_\varepsilon$ in Section 8.4, we avoided cutting off the $\log|z|$ term, which corresponds to $|z|^{4-2n}$ here, by introducing a similar term in the metric ω. This involved finding a solution of a linear equation on M_p which is asymptotic to $\log|z|$ near p. In [**103**], when constructing the approximate solution, one needs to deal in a similar way with $A_4, A_5, |z|^{2-2n}$, and $|z|^{6-4n}$. We will see shortly why better approximate solutions are useful.

The other aspect of the gluing problem is controlling the linearized operator. The operator that we need to understand is $\mathcal{D}^*_{\omega_\varepsilon}\mathcal{D}_{\omega_\varepsilon}$, where ω_ε is our approximate solution. Let us write $\mathfrak{h}_p$ for the kernel of this operator, and let $\mathfrak{h}$ denote the kernel of $\mathcal{D}^*_\omega\mathcal{D}_\omega$ on M. Note that we can naturally identify $\mathfrak{h}_p$ with a subset of $\mathfrak{h}$. If $\mathfrak{h}_p$ is non-trivial, then the best we can hope for is to invert $\mathcal{D}^*_{\omega_\varepsilon}\mathcal{D}_{\omega_\varepsilon}$ on the space orthogonal to $\mathfrak{h}_p$. In fact the operator

$$\mathcal{D}^*_{\omega_\varepsilon}\mathcal{D}_{\omega_\varepsilon} : \mathfrak{h}_p^\perp \to \mathfrak{h}_p^\perp$$

is invertible, and we need to control its inverse in suitable weighted spaces on Bl_pM. In the proof of Theorem 8.14 a crucial ingredient was the triviality of the kernel of the relevant operator on M_p, but now $\mathfrak{h}$ is contained in the kernel of $\mathcal{D}^*_\omega\mathcal{D}_\omega$, at least in the weighted spaces that we used. One could try to work in the orthogonal complement to $\mathfrak{h}$, but there will still be difficulties when $\mathfrak{h}_p$ is strictly smaller than $\mathfrak{h}$.

Remark 8.23. One could try to work with different weights δ. Note that when $\delta \in (1, 2)$, then the kernel of $\mathcal{D}^*_\omega\mathcal{D}_\omega$ in $C^{4,\alpha}_\delta(M_p)$ is just $\mathfrak{h}_p$ modulo the constants since $\mathfrak{h}_p$ consists of the elements of $\mathfrak{h}$ which vanish to first order at p. The difficulty with this is that the kernel of $\mathcal{D}^*_\eta\mathcal{D}_\eta$ in $C^{4,\alpha}_\delta(\mathrm{Bl}_0\mathbf{C}^n)$ contains functions with linear growth when $\delta > 1$.

A similar issue arises in many other problems in geometric analysis, and a way to overcome it is to rewrite our equation as a system consisting of a more general equation whose linearization we can control together with a simpler, usually finite-dimensional, equation. For our gluing problem this amounts to solving two equations of the form

$$\begin{aligned} &T_\varepsilon(\varphi, f) = 0 \quad \text{for } \varphi \in \mathfrak{h}^\perp \subset C^{4,\alpha}_\delta(\mathrm{Bl}_pM) \text{ and } f \in \mathfrak{h}, \\ &f \in \mathfrak{h}_p. \end{aligned} \tag{8.26}$$

The operator T_ε is constructed in such a way that if $T_\varepsilon(\varphi, f) = 0$ and $f \in \mathfrak{h}_p$, then $\omega_\varepsilon + \sqrt{-1}\partial\bar\partial\varphi$ is an extremal metric on Bl_pM. At the same time the linearization of T_ε is the map

$$\begin{aligned} (\mathfrak{h}^\perp \cap C^{4,\alpha}_\delta) \times \mathfrak{h} &\to C^{0,\alpha}_{\delta-4}(\mathrm{Bl}_pM) \\ (\varphi, f) &\mapsto \mathcal{D}^*_{\omega_\varepsilon}\mathcal{D}_{\omega_\varepsilon}\varphi - f, \end{aligned}$$

whose inverse one can control using the method of proof of Theorem 8.14. One technical point that we are ignoring here is that we need to make sense of the elements in $\mathfrak{h}$ as functions on Bl_pM. Since only the elements in $\mathfrak{h}_p$ have natural lifts to Bl_pM, we need to define lifts of functions in $\mathfrak{h} \setminus \mathfrak{h}_p$, but this can be achieved by using cutoff functions.

The upshot is that the equation $T_\varepsilon(\varphi, f) = 0$ can be solved for sufficiently small ε with methods very similar to the ones we used in Section 8.3. We only obtain an extremal metric if the solution satisfies $f \in \mathfrak{h}_p$, but the key observation is that we can try to vary the point $p \in M$ to achieve this. More precisely, for every $p \in M$ and for all sufficiently small ε we can solve the corresponding equation $T_\varepsilon(\varphi_{p,\varepsilon}, f_{p,\varepsilon}) = 0$, and so we have maps

$$\begin{aligned} \mu_\varepsilon : M &\to \mathfrak{h} \\ p &\mapsto f_{p,\varepsilon}. \end{aligned}$$

The construction gives rise to an extremal metric on $\mathrm{Bl}_p M$ whenever $\mu_\varepsilon(p) \in \mathfrak{h}_p$. Since $\mathfrak{h}_p \subset \mathfrak{h}$ can be thought of as vector fields vanishing at p, this condition is very reminiscent of the condition in Section 5.5 for p being a critical point of the norm squared of a moment map.

To make further progress we need to understand what the map μ_ε is, and this is where constructing a better approximate solution is important. If we construct an approximate solution that only needs to be perturbed by a term of order ε^κ to a genuine solution, then we will obtain an expression for μ_ε correct to order ε^κ. In this way the arguments in [**5**] and [**108**] lead to

$$\mu_\varepsilon(p) = \mu(p) + O(\varepsilon^\kappa)$$

for some $\kappa > 0$, where $\mu : M \to \mathfrak{h}$ is a moment map for the action of the isometry group of (M, ω). This is already enough for obtaining some existence results. For example suppose that $\mu(p) \in \mathfrak{h}_p$ for some $p \in M$. One can show that then for sufficiently small ε there is a point $q \in \mathrm{Aut}(M) \cdot p$ in the orbit of p under the automorphism group such that $\mu_\varepsilon(q) \in \mathfrak{q}$, and so we obtain an extremal metric on $\mathrm{Bl}_q M$. But $\mathrm{Bl}_q M$ is biholomorphic to $\mathrm{Bl}_p M$, so we have obtained an extremal metric on $\mathrm{Bl}_p M$. For more details on this and for sharper results, we refer the reader to [**108**] and [**103**].

8.6. K-stability of cscK manifolds

In addition to giving new examples of cscK manifolds, Theorem 8.2 also has theoretical applications. Perhaps the most important application is the following sharpening of Corollary 7.23 due to Stoppa [**101**].

Theorem 8.24. *Suppose that the compact Kähler manifold M has no holomorphic vector fields and that $\omega \in c_1(L)$ is a cscK metric. Then (M, L) is K-stable.*

Sketch of proof. Recall that a test-configuration χ for M is an embedding $M \subset \mathbf{CP}^N$ using a basis of sections of L^r for some r, together with a $\mathbf{C}^*$-action on $\mathbf{CP}^N$. We have defined the Donaldson-Futaki invariant $F(\chi)$ and the norm $\|\chi\|$, and we need to show that $F(\chi) > 0$ whenever $\|\chi\| > 0$.

We will assume that $n > 1$ since if $n = 1$, then we can consider $M \times M$ and apply Exercise 6.10. From Corollary 7.23 we already know that $F(\chi) \geqslant 0$, so let us suppose by contradiction that $F(\chi) = 0$ and $\|\chi\| > 0$. For any point $p \in M$ and large l, the line bundle $lL - E$ is an ample line bundle on the blow-up $\mathrm{Bl}_p M$, and the test-configuration χ induces a test-configuration $\hat{\chi}$ for the pair $(\mathrm{Bl}_p M, lL - E)$. The strategy of [**101**] is to choose the point $p \in M$ in such a way that for sufficiently large l we have $F(\hat{\chi}) < 0$. We know from Theorem 8.2 that $\mathrm{Bl}_p M$ admits a cscK metric in $c_1(lL - E)$ for sufficiently large l, so this contradicts Corollary 7.23.

The choice of p for this to work must be rather special since it turns out that with most choices we would actually increase the Futaki invariant of the test-configuration rather than decrease it. Suppose that the S^1-action induced by χ has Hamiltonian $h : \mathbf{CP}^N \to \mathbf{R}$. Then the action of $\chi(t)$ as $t \to 0$ along the real axis corresponds to flowing along $-\mathrm{grad}\, h$. Let $p \in M \subset \mathbf{CP}^N$ be a point where h is maximal. Then it follows that $p \in \chi(t) \cdot M$ for all t, and so $p \in M_0$, where $M_0 = \lim_{t \to 0} \chi(t) \cdot M$ as before. We will now make a simplifying assumption that M_0 is smooth in a neighborhood of p. The general case is similar but it involves a more complicated algebro-geometric calculation to deal with the possible singularity at p.

The $\mathbf{C}^*$-action χ acts on M_0, and it fixes the point $p \in M_0$, so it induces an action $\hat{\chi}$ on the blow-up $\mathrm{Bl}_p M_0$. This blow-up can be constructed locally around p using that M_0 is smooth near p. Denoting by L_0 the $\mathcal{O}(1)$-bundle restricted to M_0, we have positive line bundles $lL_0 - E$ on $\mathrm{Bl}_p M_0$ for sufficiently large l. The key calculation that we need to do is to compute the Donaldson-Futaki invariant $F(\mathrm{Bl}_p M_0, lL_0 - E, \hat{\chi})$ in terms of $F(M_0, L_0, \chi)$, where in the notation we included the line bundles $lL_0 - E$ and L_0 used for the projective embedding of $\mathrm{Bl}_0 M_0$ and M_0. We can do this, using the fact that we have an identification

$$H^0(\mathrm{Bl}_p M_0, k(lL_0 - E)) \subset H^0(M_0, klL_0)$$

of sections over $\mathrm{Bl}_p M_0$ with the sections over M_0 which vanish at p to order k. Let us denote by $\hat{A}_k$ the generator of the action $\hat{\chi}$ on $H^0(\mathrm{Bl}_p M_0, k(lL_0 - E))$ and by A_{kl} the generator of χ on $H^0(M_0, klL_0)$.

Suppose that we choose a basis

$$\{s_0, \ldots, s_{N_{kl}}\}$$

of $\mathbf{C}^*$-equivariant sections of $H^0(M_0, klL_0)$ ordered by their order of vanishing at p. So the section s_0 does not vanish, $s_1, \ldots, s_n$ vanish to first order, etc. We can use s_0 as a local trivialization of klL_0 at p, so if $z_1, \ldots, z_n$ are local holomorphic, χ-equivariant coordinates at p (using that p is a fixed point of the action), then a section s that vanishes to order r at p has an expansion of the form

$$s = f_r(z)s_0 + O(|z|^{r+1}),$$

where f_r is a degree r homogeneous polynomial. The weight $w(s)$ of the action of χ on s is given by

$$w(s) = w(s_0) + w(f_r(z)),$$

where $w(s_0) = klh(p)$ (using convention (7.11) for the action on sections), while $|w(f_r(z))| \leqslant cr$, for some c determined by the action of χ on the tangent space at p. Once l is large enough, there will be a section s corresponding to each f_r with $r \leqslant k$, so

$$\dim H^0(\mathrm{Bl}_p M_0, k(lL_0 - E)) = \dim H^0(M_0, klL_0) - \binom{k-1+n}{n}. \tag{8.27}$$

In addition we can estimate the difference between $\mathrm{Tr}(A_{kl})$ and $\mathrm{Tr}(\hat{A}_k)$ as follows:

$$\mathrm{Tr}(\hat{A}_k) = \mathrm{Tr}(A_{kl}) - \binom{k-1+n}{n} klh(p) + O(1), \tag{8.28}$$

where $O(1)$ means a term that depends on k and n but is bounded independently of l. The binomial coefficient $\binom{k-1+n}{n}$ is the dimension of the space of polynomials in $z_1, \dots, z_n$ of degree at most $k-1$.

In order to compute the Donaldson-Futaki invariants using formula (7.14), we have expansions

$$\begin{aligned}
\dim H^0(M_0, klL_0) &= a_0(kl)^n + a_1(kl)^{n-1} + \cdots, \\
\dim H^0(\mathrm{Bl}_p M_0, k(lL_0 - E)) &= \hat{a}_0 k^n + \hat{a}_1 k^{n-1} + \cdots, \\
\mathrm{Tr}(A_{kl}) &= b_0(kl)^{n+1} + b_1(kl)^n + \cdots, \\
\mathrm{Tr}(\hat{A}_k) &= \hat{b}_0 k^{n+1} + \hat{b}_1 k^n + \cdots,
\end{aligned}$$

and so from (8.27) and (8.28) we get

$$\begin{aligned}
\hat{a}_0 &= a_0 l^n - \frac{1}{n!}, \\
\hat{a}_1 &= a_1 l^{n-1} - \frac{n(n-1)}{2n!}, \\
\hat{b}_0 &= b_0 l^{n+1} - \frac{1}{n!} lh(p) + O(1), \\
\hat{b}_1 &= b_1 l^n - \frac{n(n-1)}{2n!} lh(p) + O(1),
\end{aligned}$$

where again the $O(1)$ terms are bounded independently of l. Using this we can compute

$$\begin{aligned} F(\mathrm{Bl}_p M_0, lL_0 - E, \hat{\chi}) &= \hat{b}_1 - \frac{\hat{a}_1}{\hat{a}_0}\hat{b}_0 \\ &= \left(b_1 - \frac{a_1}{a_0}b_0\right) l^n - \frac{n(n-1)}{2n!}\left(h(p) - \frac{b_0}{a_0}\right) l + O(1) \\ &= F(M_0, L_0, \chi) l^n - \frac{n(n-1)}{2n!}\left(h(p) - \frac{b_0}{a_0}\right) l + O(1). \end{aligned}$$

The assumption that $\|\chi\| > 0$ implies that h is not constant, while as in the proof of Proposition 7.15 the average of h is b_0/a_0. From the assumption that $F(\chi) = 0$ and the fact that h is maximal at p, we find that

$$F(\mathrm{Bl}_p M_0, lL_0 - E, \hat{\chi}) < 0$$

for sufficiently large l. This gives the contradiction that we wanted. □

Bibliography

1. M. Abreu, *Kähler geometry of toric varieties and extremal metrics*, Internat. J. Math. **9** (1998), 641–651.

2. V. Apostolov, D. M. J. Calderbank, P. Gauduchon, and C. W. Tønnesen-Friedman, *Hamiltonian 2-forms in Kähler geometry III, extremal metrics and stability*, Invent. Math. **173** (2008), no. 3, 547–601.

3. C. Arezzo and F. Pacard, *Blowing up and desingularizing constant scalar curvature Kähler manifolds*, Acta Math. **196** (2006), no. 2, 179–228.

4. ______, *Blowing up Kähler manifolds with constant scalar curvature II*, Ann. of Math. (2) **170** (2009), no. 2, 685–738.

5. C. Arezzo, F. Pacard, and M. A. Singer, *Extremal metrics on blow ups*, Duke Math. J. **157** (2011), no. 1, 1–51.

6. M. F. Atiyah, *Convexity and commuting Hamiltonians*, Bull. London Math. Soc. **14** (1982), no. 1, 1–15.

7. T. Aubin, *Métriques riemanniennes et courbure*, J. Differential Geom. **4** (1970), 383–424.

8. ______, *Équations du type Monge-Ampère sur les varietés kählériennes compactes*, Bull. Sci. Math. (2) **102** (1978), no. 1, 63–95.

9. ______, *Réduction de cas positif de l'équation de Monge-Ampère sur les variétés kählériennes compactes à la démonstration d'une inégalité*, J. Funct. Anal. **57** (1984), no. 2, 143–153.

10. ______, *Some nonlinear problems in Riemannian geometry*, Springer Monographs in Mathematics, Springer-Verlag, Berlin, 1998.

11. S. Bando and T. Mabuchi, *Uniqueness of Einstein Kähler metrics modulo connected group actions*, Algebraic geometry, Sendai, Adv. Stud. Pure Math., vol. 10, 1985, pp. 11–40.

12. R. Bartnik, *The mass of an asymptotically flat manifold*, Comm. Pure Appl. Math. **39** (1986), no. 5, 661–693.

13. E. Bedford and B. A. Taylor, *The Dirichlet problem for a complex Monge-Ampère equation*, Invent. Math. **37** (1976), no. 1, 1–44.

14. R. Berman, *K-polystability of Q-Fano varieties admitting Kähler-Einstein metrics*, arXiv:1205:6214v2.

15. B. Berndtsson, *A Brunn-Minkowski type inequality for Fano manifolds and some uniqueness theorems in Kähler geometry*, arXiv:1303.4975.

16. O. Biquard and Y. Rollin, *Smoothing singular extremal Kähler surfaces and minimal lagrangians*, arXiv:1211.6957.

17. Z. Błocki, *The Calabi-Yau theorem*, Complex Monge-Ampère equations and geodesics in the space of Kähler metrics, Lecture Notes in Math., vol. 2038, Springer, Heidelberg, 2012, pp. 201–227.

18. S. Boucksom and H. Chen, *Okounkov bodies of filtered linear series*, Compos. Math. **147** (2011), 1205–1229.

19. S. Boucksom, P. Eyssidieux, V. Guedj, and A. Zeriahi, *Monge-Ampère equations in big cohomology classes*, Acta Math. **205** (2010), no. 2, 199–262.

20. L. Caffarelli, J. J. Kohn, L. Nirenberg, and J. Spruck, *The Dirichlet problem for nonlinear second-order elliptic equations II. Complex Monge-Ampère and uniformly elliptic equations*, Comm. Pure Appl. Math. **38** (1985), 209–252.

21. E. Calabi, *Extremal Kähler metrics*, Seminar on Differential Geometry (S. T. Yau, ed.), Princeton, 1982.

22. ———, *Extremal Kähler metrics II*, Differential geometry and complex analysis, Springer, 1985, pp. 95–114.

23. H. D. Cao, *Deformation of Kähler metrics to Kähler-Einstein metrics on compact Kähler manifolds*, Invent. Math. **81** (1985), no. 2, 359–372.

24. D. Catlin, *The Bergman kernel and a theorem of Tian*, Analysis and geometry in several complex variables (Katata, 1997), Trends Math., Birkhäuser Boston Inc., 1999, pp. 1–23.

25. J. Cheeger and T. Colding, *On the structure of spaces with Ricci curvature bounded below. I*, J. Differential Geom. **46** (1997), no. 3, 406–480.

26. ———, *On the structure of spaces with Ricci curvature bounded below. II*, J. Differential Geom. **54** (2000), no. 1, 13–35.

27. ———, *On the structure of spaces with Ricci curvature bounded below. III*, J. Differential Geom. **54** (2000), no. 1, 37–74.

28. B. Chen, An-Min Li, and L. Sheng, *Extremal metrics on toric surfaces*, arXiv: 1008.2607.

29. X. X. Chen, *The space of Kähler metrics*, J. Differential Geom. **56** (2000), no. 2, 189–234.

30. X. X. Chen, *Space of Kähler metrics III. On the lower bound of the Calabi energy and geodesic distance*, Invent. Math. **175** (2009), no. 3, 453–503.

31. X. X. Chen, S. K. Donaldson, and S. Sun, *Kähler-Einstein metrics and stability*, arXiv:1210.7494.

32. ———, *Kähler-Einstein metrics on Fano manifolds, I: Approximation of metrics with cone singularities*, arXiv:1211.4566.

33. ———, *Kähler-Einstein metrics on Fano manifolds, II: Limits with cone angle less than 2π*, arXiv:1212.4714.

34. ———, *Kähler-Einstein metrics on Fano manifolds, III: Limits as cone angle approaches 2π and completion of the main proof*, arXiv:1302.0282.

35. X. X. Chen, C. LeBrun, and B. Weber, *On conformally Kähler, Einstein manifolds*, J. Amer. Math. Soc. **21** (2008), no. 4, 1137–1168.

36. X. X. Chen and G. Tian, *Geometry of Kähler metrics and foliations by holomorphic discs*, Publ. Math. Inst. Hautes Études Sci. (2008), no. 107, 1–107.

37. P. Cherrier, *Équations de Monge-Ampère sur les variétés Hermitiennes compactes*, Bull. Sci. Math. (2) **111** (1987), no. 4, 343–385.

38. T. Darvas, *Morse theory and geodesics in the space of Kähler metrics*, arXiv: 1207.4465.

39. J.-P. Demailly, *Complex analytic and differential geometry*, `http://www-fourier.ujf-grenoble.fr/~demailly/manuscripts/agbook.pdf`.

40. W. Y. Ding, *Remarks on the existence problem of positive Kähler-Einstein metrics*, Math. Ann. **282** (1988), no. 3, 463–471.

41. S. K. Donaldson, *Lecture notes for TCC course "Geometric Analysis"*, `http://www2.imperial.ac.uk/~skdona/GEOMETRICANALYSIS.PDF`.

42. ______, *Remarks on gauge theory, complex geometry and four-manifold topology*, Fields Medallists' Lectures (Atiyah and Iagolnitzer, eds.), World Scientific, 1997, pp. 384–403.

43. ______, *Symmetric spaces, Kähler geometry and Hamiltonian dynamics*, Northern California Symplectic Geometry Seminar, Amer. Math. Soc. Transl. Ser. 2, vol. 196, Amer. Math. Soc., Providence, RI, 1999, pp. 13–33.

44. ______, *Scalar curvature and stability of toric varieties*, J. Differential Geom. **62** (2002), 289–349.

45. ______, *Interior estimates for solutions of Abreu's equation*, Collect. Math. **56** (2005), no. 2, 103–142.

46. ______, *Lower bounds on the Calabi functional*, J. Differential Geom. **70** (2005), no. 3, 453–472.

47. ______, *Two-forms on four-manifolds and elliptic equations*, inspired by S. S. Chern, Nankai Tracts Math., vol. 11, World Sci. Publ., Hackensack, NJ, 2006, pp. 153–172.

48. ______, *Extremal metrics on toric surfaces: A continuity method*, J. Differential Geom. **79** (2008), no. 3, 389–432.

49. ______, *Constant scalar curvature metrics on toric surfaces*, Geom. Funct. Anal. **19** (2009), no. 1, 83–136.

50. ______, *Discussion of the Kähler Einstein problem*, `http://www2.imperial.ac.uk/~skdona/KENOTES.PDF` (2009).

51. S. K. Donaldson and S. Sun, *Gromov-Hausdorff limits of Kähler manifolds and algebraic geometry*, arXiv:1206.2609.

52. P. Eyssidieux, V. Guedj, and A. Zeriahi, *Singular Kähler-Einstein metrics*, J. Amer. Math. Soc. **22** (2009), no. 3, 607–639.

53. A. Fujiki, *Moduli space of polarized algebraic manifolds and Kähler metrics [translation of Sūgaku* **42** *(1990), no. 3, 231–243;]*, Sugaku Expositions **5** (1992), no. 2, 173–191.

54. W. Fulton, *Introduction to toric varieties*, Ann. of Math. Stud., vol. 131, Princeton University Press, Princeton, N.J., 1993.

55. A. Futaki, *An obstruction to the existence of Einstein-Kähler metrics*, Invent. Math. **73** (1983), 437–443.

56. A. Futaki and T. Mabuchi, *Bilinear forms and extremal Kähler vector fields associated with Kähler classes*, Math. Ann. **301** (1995), 199–210.

57. P. Gauduchon, *Invariant scalar-flat Kähler metrics on $\mathcal{O}(-l)$*, preprint, 2012.

58. D. Gilbarg and N. S. Trudinger, *Elliptic partial differential equations of second order*, reprint of the 1998 ed., Classics in Mathematics, Springer-Verlag, Berlin, 2001.

59. P. Griffiths and J. Harris, *Principles of algebraic geometry*, Pure and Applied Mathematics, Wiley-Interscience, New York, 1978.

60. D. Guan, *On modified Mabuchi functional and Mabuchi moduli space of Kähler metrics on toric bundles*, Math. Res. Lett. **6** (1999), no. 5-6, 547–555.

61. V. Guillemin, *Kaehler structures on toric varieties*, J. Differential Geom. **40** (1994), 285–309.

62. V. Guillemin and S. Sternberg, *Convexity properties of the moment mapping*, Invent. Math. **67** (1982), no. 3, 491–513.

63. ———, *Riemann sums over polytopes*, Ann. Inst. Fourier (Grenoble) **57** (2007), no. 7, 2183–2195.

64. R. Hamilton, *Three-manifolds with positive Ricci curvature*, J. Differential Geom. **17** (1982), no. 2, 255–306.

65. A. Hwang and M. A. Singer, *A momentum construction for circle-invariant Kähler metrics*, Trans. Amer. Math. Soc. **354** (2002), no. 6, 2285–2325.

66. W. Jiang, *Bergman kernel along the Kähler Ricci flow and Tian's conjecture*, arXiv:1311.0428.

67. F. C. Kirwan, *Cohomology of quotients in symplectic and algebraic geometry*, Princeton University Press, 1984.

68. S. Kołodziej, *The complex Monge-Ampère equation*, Acta Math. **180** (1998), no. 1, 69–117.

69. ———, *Hölder continuity of solutions to the complex Monge-Ampère equation with the right-hand side in L^p: The case of compact Kähler manifolds*, Math. Ann. **342** (2008), no. 2, 379–386.

70. C. LeBrun, *Counter-examples to the generalized positive action conjecture*, Comm. Math. Phys. **118** (1988), no. 4, 591–596.

71. C. LeBrun and S. R. Simanca, *On the Kähler classes of extremal metrics*, Geometry and global analysis (Sendai, 1993), Tohoku Univ., 1993, pp. 255–271.

72. ———, *Extremal Kähler metrics and complex deformation theory*, Geom. and Func. Anal. **4** (1994), no. 3, 298–336.

73. L. Lempert and L. Vivas, *Geodesics in the space of Kähler metrics*, Duke Math. J. **162** (2013), no. 7, 1369–1381.

74. M. Levine, *A remark on extremal Kähler metrics*, J. Differential Geom. **21** (1985), no. 1, 73–77.

75. C. Li and C. Xu, *Special test configurations and K-stability of Fano varieties*, arXiv:1111.5398.

76. A. Lichnerowicz, *Sur les transformations analytiques des variétés kählériennes compactes*, C. R. Acad. Sci. Paris **244** (1957), 3011–3013.

77. R. B. Lockhart and R. C. McOwen, *Elliptic differential operators on noncompact manifolds*, Ann. Sc. Norm. Super. Pisa Cl. Sci. (4) **1** (1985), no. 3, 409–447.

78. Z. Lu, *On the lower order terms of the asymptotic expansion of Tian-Yau-Zelditch*, Amer. J. Math. **122** (1998), no. 2, 235–273.

79. T. Mabuchi, *K-stability of constant scalar curvature polarization*, arXiv:0812.4093.

80. ———, *K-energy maps integrating Futaki invariants*, Tohoku Math. J. **38** (1986), no. 4, 575–593.

81. ______, *Some symplectic geometry on compact Kähler manifolds*, Osaka J. Math. **24** (1987), 227–252.

82. M. Maruyama, *On automorphism groups of ruled surfaces*, J. Math. Kyoto University **11-1** (1971), 89–112.

83. Y. Matsushima, *Sur la structure du groupe d'homéomorphismes analytiques d'une certaine variété kählérienne*, Nagoya Math. J. **11** (1957), 145–150.

84. D. McDuff and D. Salamon, *Introduction to symplectic topology*, OUP, 1998.

85. D. Mumford, J. Fogarty, and F. Kirwan, *Geometric invariant theory*, third ed., Ergebnisse der Mathematik und ihrer Grenzgebiete (2) [Results in Mathematics and Related Areas (2)], vol. 34, Springer-Verlag, Berlin, 1994. MR1304906 (95m:14012)

86. F. Pacard, *Lectures on connected sum constructions in geometry and nonlinear analysis*, `http://www.math.polytechnique.fr/~pacard/Publications/Lecture-Part-I.pdf`.

87. S. T. Paul, *Stable pairs and coercive estimates for the Mabuchi functional*, arXiv:1308.4377.

88. S. T. Paul and G. Tian, *CM stability and the generalised Futaki invariant II*, Astérisque **328** (2009), 339–354.

89. G. Perelman, *The entropy formula for the Ricci flow and its geometric applications*, math.DG/0211159.

90. D. H. Phong, N. Sesum, and J. Sturm, *Multiplier ideal sheaves and the Kähler-Ricci flow*, Comm. Anal. Geom. **15** (2007), no. 3, 613–632.

91. D. H. Phong, J. Song, and J. Sturm, *Complex Monge-Ampère equations*, Surveys in differential geometry, vol. 17, Int. Press, Boston, MA, 2012, pp. 327–410.

92. J. Ross and R. P. Thomas, *A study of the Hilbert-Mumford criterion for the stability of projective varieties*, J. Algebraic Geom. **16** (2007), no. 2, 201–255.

93. W.-D. Ruan, *Canonical coordinates and Bergman metrics*, Comm. Anal. Geom. **6** (1998), no. 3, 589–631.

94. S. Semmes, *Complex Monge-Ampère equations and symplectic manifolds*, Amer. J. Math. **114** (1992), 495–550.

95. S. R. Simanca, *Kähler metrics of constant scalar curvature on bundles over $\mathbb{C}\mathrm{P}_{n-1}$*, Math. Ann. **291** (1991), no. 2, 239–246.

96. L. Simon, *Theorems on regularity and singularity of energy minimizing maps*, Lectures in Mathematics ETH Zürich, Birkhäuser Verlag, Basel, 1996.

97. Y. T. Siu, *Lectures on Hermitian-Einstein metrics for stable bundles and Kähler-Einstein metrics*, DMV Seminar 8, Birkhäuser Verlag, Basel, 1987.

98. J. Song and G. Tian, *The Kähler-Ricci flow on surfaces of positive Kodaira dimension*, Invent. Math. **170** (2007), no. 3, 609–653.

99. ______, *Canonical measures and Kähler-Ricci flow*, J. Amer. Math. Soc. **25** (2012), no. 2, 303–353.

100. J. Song and B. Weinkove, *Lecture notes on the Kähler-Ricci flow*, arXiv:1212.3653.

101. J. Stoppa, *K-stability of constant scalar curvature Kähler manifolds*, Adv. Math. **221** (2009), no. 4, 1397–1408.

102. J. Stoppa and G. Székelyhidi, *Relative K-stability of extremal metrics*, J. Eur. Math. Soc. **13** (2011), no. 4, 899–909.

103. G. Székelyhidi, *Blowing up extremal Kähler manifolds II*, arXiv:1302.0760.

104. ———, *Filtrations and test-configurations, with an appendix by S. Boucksom*, arXiv:1111.4986.

105. ———, *The partial C^0-estimate along the continuity method*, arXiv:1310.8471.

106. ———, *Extremal metrics and K-stability*, Ph.D. thesis, Imperial College, London, 2006.

107. ———, *Extremal metrics and K-stability*, Bull. Lond. Math. Soc. **39** (2007), no. 1, 76–84.

108. ———, *On blowing up extremal Kähler manifolds*, Duke Math. J. **161** (2012), no. 8, 1411–1453.

109. R. P. Thomas, *Notes on GIT and symplectic reduction for bundles and varieties*, Surveys in Differential Geometry **10** (2006): A Tribute to Professor S.-S. Chern.

110. G. Tian, *On a set of polarized Kähler metrics on algebraic manifolds*, J. Differential Geom. **32** (1990), no. 1, 99–130.

111. ———, *On Calabi's conjecture for complex surfaces with positive first Chern class*, Invent. Math. **101** (1990), no. 1, 101–172.

112. ———, *Kähler-Einstein metrics with positive scalar curvature*, Invent. Math. **137** (1997), 1–37.

113. ———, *Canonical metrics in Kähler geometry*, Lectures in Mathematics ETH Zürich, Birkhäuser Verlag, Basel, 2000.

114. ———, *Existence of Einstein metrics on Fano manifolds*, Metric and Differential Geometry, The Jeff Cheeger Anniversary Volume (X. Dai and X. Rong, eds.), Progress in Mathematics, vol. 297, Springer, Basel, 2012, pp. 119–159.

115. G. Tian and Z. Zhang, *Regularity of Kähler-Ricci flows on Fano manifolds*, arXiv:1310.5897.

116. C. W. Tønnesen-Friedman, *Extremal Kähler metrics on minimal ruled surfaces*, J. Reine Angew. Math. **502** (1998), 175–197.

117. V. Tosatti and B. Weinkove, *The complex Monge-Ampère equation on compact hermitian manifolds*, J. Amer. Math. Soc. **23** (2010), no. 4, 1187–1195.

118. ———, *The Calabi-Yau equation, symplectic forms and almost complex structures*, Geometry and analysis. No. 1, Adv. Lect. Math. (ALM), vol. 17, Int. Press, Somerville, MA, 2011, pp. 475–493.

119. X. Wang, *Moment map, Futaki invariant and stability of projective manifolds*, Comm. Anal. Geom. **12** (2004), no. 5, 1009–1037.

120. X.-J. Wang and B. Zhou, *On the existence and nonexistence of extremal metrics on toric Kähler surfaces*, Adv. Math. **226** (2011), no. 5, 4429–4455.

121. D. Witt Nyström, *Test configurations and Okounkov bodies*, Compos. Math. **148** (2012), no. 6, 1736–1756.

122. S.-T. Yau, *On the Ricci curvature of a compact Kähler manifold and the complex Monge-Ampère equation I*, Comm. Pure Appl. Math. **31** (1978), 339–411.

123. ———, *Open problems in geometry*, Proc. Symposia Pure Math. **54** (1993), 1–28.

124. S. Zelditch, *Szegő kernel and a theorem of Tian*, Int. Math. Res. Notices **6** (1998), 317–331.

125. ———, *Bernstein polynomials, Bergman kernels and toric Kähler varieties*, J. Symplectic Geom. **7** (2009), no. 2, 51–76.

Index